PENGUIN BOOKS

BEFORE THE DAWN

Nicholas Wade is a reporter at the *New York Times*. He previously worked for *Nature* and *Science*, two leading scientific journals, and has written five previous books.

D0029354

BEFORE THE DAWN

Recovering

the Lost History *of*

Our Ancestors

NICHOLAS WADE

PENGUIN BOOKS

PENGUIN BOOKS

Published by the Penguin Group

Penguin Group (USA) Inc., 375 Hudson Street, New York, New York 10014, U.S.A.
Penguin Group (Canada), 90 Eglinton Avenue East, Suite 700, Toronto,
Ontario, Canada M4P 2Y3 (a division of Pearson Penguin Canada Inc.)
Penguin Books Ltd, 80 Strand, London WC2R 0RL, England
Penguin Ireland, 25 St Stephen's Green, Dublin 2, Ireland (a division of Penguin Books Ltd)
Penguin Group (Australia), 250 Camberwell Road, Camberwell,
Victoria 3124, Australia (a division of Pearson Australia Group Pty Ltd)
Penguin Books India Pvt Ltd, 11 Community Centre,
Panchsheel Park, New Delhi–110 017, India
Penguin Group (NZ), 67 Apollo Drive, Mairangi Bay, Auckland 1311,
New Zealand (a division of Pearson New Zealand Ltd)
Penguin Books (South Africa) (Pty) Ltd, 24 Sturdee Avenue, Rosebank,
Johannesburg 2196, South Africa

Penguin Books Ltd, Registered Offices: 80 Strand, London WC2R 0RL, England

First published in the United States of America by The Penguin Press,
a member of Penguin Group (USA) Inc. 2006
Published in Penguin Books 2007

7 9 10 8

Illustration credits appear on pages 301–302.

THE LIBRARY OF CONGRESS HAS CATALOGED THE HARDCOVER
EDITION AS FOLLOWS:
Wade, Nicholas.
Before the dawn : recovering the lost history of our ancestors / Nicholas Wade.
p.   cm.
Includes bibliographical references and index.
ISBN 1-59420-079-3 (hc.)
ISBN 978-0-14-303832-0 (pbk.)
1. Human evolution.   2. Social evolution.   I. Title.
GN281.W33 2006
599.93'8—dc22   2005055293

Printed in the United States of America
Designed by Kate Nichols

Contents

1

Genetics & Genesis

It has often and confidently been asserted, that man's origin can never be known: but ignorance more frequently begets confidence than does knowledge: it is those who know little, and not those who know much, who so positively assert that this or that problem will never be solved by science.

CHARLES DARWIN, THE DESCENT OF MAN

TRAVEL BACK INTO THE HUMAN PAST, and the historical evidence is plentiful enough for the first couple of hundred years, then rapidly diminishes. At the 5,000-year mark written records disappear altogether, yielding to the wordless witness of archaeological sites. Going farther back, even these become increasingly rare over the next 10,000 years, fading almost to nothing by 15,000 years ago, the date of the first human settlements. Before that time, people lived a nomadic existence based on hunting and gathering. They built nothing and left behind almost nothing of permanence, save a few stone tools and the remarkable painted caves of Europe.

Travel on back for another 35,000 years and you will have reached the 50,000-year mark, the time when the ancestral human population was still confined to its homeland somewhere in northeast Africa but had begun to show the first signs of modern behavior. If this is the point at which the modern human story begins, then written records exist for just the last 10% of it; 90% of human history seems irretrievably lost.

Keep traveling back in time to the earliest starting point in the human narrative, the period 5 million years ago when the ape-like creatures at the head of the human line of descent split from those at the head of the chimpanzee line of descent. The only physical evidence from throughout this period, which saw the evolution from ape to human form, is a handful of battered skulls and a few stone tools.

No deep understanding, it might seem, could ever be gained of these two vanished periods, the 5 million years of human evolution and the 45,000 years of prehistory. But in the past few years an extraordinary new archive has become available to those who study human evolution, human nature and history. It is the record encoded in the DNA of the human genome and in the versions of it carried by the world's population. Geneticists have long contributed to the study of the human past but are doing so with particular success since the full sequence of DNA units in the genome was determined in 2003.

Why should the human genome, specifically shaped for survival in the present, have so much to say about the past? As the repository of hereditary information that is in constant flux, the genome is like a document under ceaseless revision. Its mechanism of change is such that it retains evidence about its previous drafts and these, though not easy to interpret, provide a record that stretches deep into the past. The genome can therefore be interrogated at many different time levels. It can supply answers that reach back more than 50,000 years to the genetic Adam, a man whose Y chromosome is carried by all men now alive. Or it can be queried about the events of a mere couple of centuries ago, such as whether Thomas Jefferson, the third president of the United States, had a secret family with his slave mistress Sally Hemings.

From Adam to Jefferson, the genome is helping researchers create a new and far more detailed picture of human evolution, human nature and history. From the great darkness, a surprisingly full narrative is emerging. This new narrative of the human past rests on a solid foundation laid by paleoanthropologists, archaeologists, anthropologists and many other specialists. It can be called new in the sense that genetic information now contributes to each of these traditional disciplines and is beginning to draw them together.

This book describes those aspects of human evolution, nature and prehistory that have been illumined by genetic discoveries of the last few years. Readers who do not follow these fields closely may be surprised at the richness of the information in the new narrative. There exists no video of how apes slowly morphed into people, but a sequence of the salient events can for the most part be reconstructed. There is no map that records the dispersal of the new humans from their ancestral homeland, but researchers can now follow the path they took out of Africa and their migrations through the world outside. It's even possible to reconstruct some of the social institutions that emerged as people made the transition from a nomadic way of life, based on hunting and gathering, to today's complex societies.

Information from the genome has helped tell paleoanthropologists when humans lost their body hair and when they gained the power of speech. It has clarified for archaeologists their long quandary as to whether Neanderthals and modern humans peacefully interbred with each other or fought until the Neanderthals' extinction. It has furnished anthropology with information about human adaptation to cultural practices like cattle-herding and cannibalism. The cascade of DNA data is even benefiting historical linguistics, though indirectly, as biologists apply the tree-building methods developed for gene genealogies to reconstructing the evolution of language.

On the critical question of the ancestral human population of 50,000 years ago, the last group from which everyone alive today is descended, the techniques of paleoanthropology and archaeology are powerless to say anything about a people that has vanished without trace. But geneticists, by rummaging around in the genome's rich attic, can fill in all kinds of unexpected detail. They can estimate how large the ancestral population was. They can say where in Africa it probably lived. They can put a date, though a rough one, on when language emerged. They can even infer, in one instance, what the first language sounded like.

The First Tailored Clothes

Few findings better illustrate geneticists' ability to cast light into surprising corners of the human past than a recent estimate of the date that people first sewed their own clothes. Early humans may have used loose animal skins for millions of years, worn perhaps like a cape against the cold, but fabricated garments were a more recent invention. Archaeologists have never been able to determine when clothes were first worn because both the materials and the bone needles used to sew them are highly perishable.

In the fall of 1999, Mark Stoneking's son came home from school in Leipzig, Germany, with a note warning that a classmate had a case of head lice. Stoneking, an American researcher at the Max Planck Institute for Evolutionary Anthropology, read it as carefully as would any anxious parent. But as a geneticist long interested in human origins, his attention was drawn to a reassurance in the school note that lice cannot survive longer than 24 hours away from the warmth of the human body. "I thought if that was true, then lice must have been spread around the world by human migrations," he says. Stoneking figured that if he could prove this were so, he would have discovered an independent confirmation of the migration pattern implied by human DNA. But after a few hours of library research, he realized that the lice might hold in their DNA an even more interesting fact—the date when humans first wore clothing.

The compilers of the book of Genesis were so exercised by the question of human nakedness that they included not just one but two accounts of how people came to seek modesty in clothing. In the first, Adam and Eve sewed themselves aprons of fig leaves after realizing their state of undress. In the second, the Creator himself tailored the errant pair coats of skins before expelling them into the world beyond Eden.[1] Neither account gives due weight to the other interested party in the story of human clothing, the louse. Once, after all, in the days when human forebears were fully covered with hair just like any other ape or monkey, the louse must have ranged freely from head to toe.

When humans lost their body hair, the louse's domain shrank, confining it to the lonely island of hair that tufts absurdly from the human head. But it patiently bided its time and many millennia later, when people started to wear clothes, the head louse seized the chance to regain its lost territory by evolving a new variety, the body louse, that could live in clothing. The head and body louse closely resemble each other except the body louse is larger and has claws specialized for grasping material, not shafts of hair. Stoneking realized that he could date the invention of clothing if he could only figure out from variations in lice DNA the time at which the body louse began to evolve from the head louse.

He collected head and body lice from the citizens of 12 countries around the world, from Ethiopia to Ecuador to New Guinea. He analyzed all the variations in a small segment of each head and body louse's genetic material and arranged each population's lice in a family tree. Knowing the rate at which variations accumulate on DNA over the centuries, he could then cal-culate the dates of the various forks or branch points in the tree.

The branch point at which the body louse first evolved from the head louse turned out to be around 72,000 years ago, give or take several thousand years either way.[2] Assuming the body louse evolved almost immediately after its new niche was available to it, then people first addressed their nakedness only in the most recent stage of their evolutionary history. It was about this time, or a few thousand years later, that people perfected language and broke out of Africa to colonize the rest of the world. It seems they had decided to get dressed for the occasion.

From Adam to Jefferson

Genetics, with its fresh new insights into the human past, ranges across many other academic territories. At least seven traditional disciplines bear on the human past. *Paleoanthropologists*, the students of fossil human remains, have reconstructed the major steps by which the human lineage branched off from apes 5 million years ago and, by 100,000 years ago, had morphed into humans who were anatomically though not behaviorally similar to people of today. *Archaeologists* have picked up the story from there, establishing the foundation of dates and basic facts on which other specialists seek to reconstruct various aspects of past human behavior. *Population geneticists* have tracked the migration of human populations around the world. Their early analyses were based on differences in human proteins but the emphasis has now switched to DNA, a more convenient and informative source.

Historical linguists have traced back the family tree of human languages, reconstructing vanished tongues such as proto-Indo-European, the inferred ancestral tongue of many languages spoken in Europe, Iran and India. *Primatologists*, after many years of patient observation, have gained a deep understanding of how chimpanzee and bonobo societies work. This achievement provides insights into the social organization of the primates from which both chimps and people evolved, since chimps may closely resemble those ancient ancestors. *Social anthropologists*, through the study of surviving hunter-gatherers and other primitive societies, have laid the basis for reconstructing the evolution of human social structures. *Evolutionary psychologists* seek to identify the tasks that evolution has designed the mind to perform. In two other fields closely related to evolutionary psychology, those of human behavioral ecology and evolutionary anthropology, researchers explore ways of applying the principles of evolutionary biology to human societies. From these three subjects have emerged many sharp insights into how the search for reproductive advantage shapes people's choices in marriage, parenting and the allocation of their resources.

Researchers in each of these seven disciplines have helped delineate the distant human past, often by ingenious interpretation of fragmentary evidence. The seven traditional disciplines are increasingly being aided by an eighth, that of evolutionary biology, the body of theory on which evolutionary psychology seeks to draw. Many specialists have assumed that evolutionary change works so slowly that its effect on the recent human past, if any, can safely be ignored. But it was only lack of knowledge that made it seem evolution's hand had been stayed. As is now evident from analyses of DNA, human genes have continued to evolve until the present day. Like everything else in biology, the human past and present are incomprehensible except in the light of evolution.

The human genome is a new source of data that enriches all the disciplines concerned with the human past. It furnishes two quite different types of information, one to do with genes, the other with genealogies.

New versions of existing genes often arise in the course of evolution, and become more common in a population because they confer some advantage. These new versions carry distinctive properties that allow geneticists to estimate their approximate age. So when a gene is found that concerns some major feature of human evolution, like the FOXP2 gene, which is involved in language, or the melanocortin receptor gene which influences skin color, geneticists can often set dates on the window of time in which the feature evolved.

A second kind of information in the genome allows ancestries to be traced, usually through a special part of the genome like the Y chromosome, which is passed down essentially unchanged from one generation to another. Every few generations a mutation—the random conversion of one DNA unit to another—occurs on the Y, with the result that all descendants of the man in whom the mutation occurred will also carry it. All men can be assigned to different lineages, based on the particular pattern of mutations they carry on their Y chromosome. These patterns allow many inferences to be drawn about human migrations because the lineages for the most part are confined to the specific geographical regions where their owners first settled.

The human genome thus records a vast span of the human past and enriches the findings of traditional disciplines. Following are the principal themes, explored in the pages that follow, to which DNA has added new insights:

• *There is a clear continuity between the ape world of 5 million years ago and the human world that emerged from it.* The thread is most visible at the level of DNA: the genomes of humans and chimpanzees are 99% identical. It is evident enough in the physical resemblance between the two species. But perhaps the most interesting level of continuity is between the social institutions of the ape and human worlds.

The apes ancestral to both chimpanzees and humans probably lived as small bands of related individuals who defended a home territory, often with lethal attacks against neighbors. They had separate male and female hierarchies and most infants were sired by the society's dominant male or his allies. The emerging human line was also territorial but in time developed a new social structure based on pair bonding, a stable relationship between a male and one or more females. This critical shift would have given all males a chance of reproduction and hence a stronger interest in the group's welfare, making human societies larger and more cohesive.

• *A principal force in the shaping of human evolution has been the nature of human society.* After splitting from the apes, those in the human line of descent evolved upright stature and developed dark skin in place of the ape's body hair. But the most significant change—a steady increase in brain size—probably evolved in response to the most critical aspect of the environment, the society in which an individual lived. Judging whom to trust, forming alliances, keeping score of favors given and received—all were necessities made easier by greater cognitive ability. By 50,000 years ago, the social benefits of more efficient communication had prompted the evolution of a novel ability possessed by no other social species, the faculty of language.

• *The human physical form was attained first, followed by continued evolution of human behavior.* Anatomically modern humans, people whose physical remains resemble the skeletons of people today, became common 100,000 years ago. But they showed no sign of the advanced behaviors that emerged 50,000 years later, probably made possible by the evolution of language. With this new faculty and the greater social cohesion it provided, the first behaviorally modern humans were able to break out of Africa and displace the archaic humans like the Neanderthals who had left Africa many thousands of years previously.

• *Most of human prehistory occurred in, and was shaped by, the last ice age.* The first modern humans to leave Africa probably crossed over the Red Sea at its southern end and into Arabia. Reaching India, the population went separate ways. One group traveled along the coasts of southeast Asia, arriving in Australia some 46,000 years ago. Another explored the land route northwest from India, reaching Europe and slowly evicting the Neanderthals from their ancient homeland. The expansion into the cold northern latitudes of Eurasia required technical innovation and probably genetic adaptations too. Then a climatic catastrophe, the return of the glaciers 20,000 years ago, emptied Europe and Siberia of people. Descendants of the survivors spread north again several thousand years later as the Pleistocene ice age drew to a close. Some of these new northerners, the Siberians in the eastern half of Eurasia, contrived the first domestication, that of the dog, and discovered the land bridge that then joined Siberia to Alaska and the Americas.

• *The adaptations for three principal social institutions, warfare, religion and trade, had evolved by 50,000 years ago.* The ancestral human population, the first to possess the power of fully articulate modern speech, may have numbered only 5,000 people, confined to a homeland in northeast Africa. These ancestral people, though less cognitively advanced than people today, possessed all the distinctive features of human nature and had developed, at least in rudimentary form, the institutions that are found in societies throughout the world. These may have included warfare centered round a defense of territory, religious ceremony as a means of social cohesion, and an instinct for reciprocity that governed social relations within the group and trade with those outside it.

• *The ancestral people had a major limitation to overcome: they were too aggressive to live in settled communities.* Early human societies lived as small bands of hunter-gatherers, their existence dominated by incessant warfare. For 35,000 years after leaving the ancestral homeland, these nomads were unable to settle down. Only gradually did humans evolve to become less aggressive. The tempo of warfare eased and a more gracile, or delicately boned, human form evolved in populations throughout the world. In the Near East, around 15,000 years ago, people at last accomplished a decisive social transition, the founding of the first settled communities. In place of the hunter-gatherers' egalitarianism and lack of possessions, people in settled societies developed a new social structure with elites, specialization of roles, and ownership of property. Human groups started for the first time to produce storable surpluses of food and other products, which led to more complex societies and to increased trade between groups.

• *Human evolution did not halt in the distant past but has continued to the present day.* The ancestral human population of 50,000 years ago differed greatly from the anatomically modern humans of 100,000 years ago, and people today have had just as long to evolve away from the ancestral population. The human genome bears many marks of recent evolution, prompted by adaptations to events such as cultural changes or new diseases.

More visible evidence of recent evolution is the existence of human races. After the dispersal of the ancestral population from Africa 50,000 years ago, human evolution continued independently in each continent. The populations of the world's major geographical regions bred for many thousand years in substantial isolation from each other and started to develop distinctive features, a genetic differentiation which is the basis for today's races. But these separate evolutionary paths were to some extent parallel as people in different continents responded to the same challenges. Gracilization occurred worldwide. Lactose tolerance, the genetic ability to digest lactose in adulthood, evolved among cattle-herding people in Europe 5,000 years ago but also among pastoral peoples of Africa and the Middle East.

• *People probably once spoke a single language from which all contemporary languages are derived.* Just as the ancestral population, after its dispersal,

diverged into different races and ethnic groups, the ancestral tongue split
into a growing family of different languages. Some of these languages expanded under the influence of factors such as warfare or agriculture, so
that certain language families, like Indo-European, are now spoken over
large areas while others, like many in South America or New Guinea, have
a range of a few miles. Because language splits follow population splits,
the genealogy of human languages must mirror, to some extent, the tree
of descent of human populations; some biologists hope that the genealogy of human languages can be reconstructed far into the past, perhaps
even near to its root, the mother tongue of the ancestral human population.

• *The human genome contains excellent records of the recent past, providing
a parallel history to the written record.* The genome evolves so fast that
whenever any community starts to breed in isolation, whether for reasons
of religion, geography or language, within a few centuries its genetics assume a distinctive signature. DNA sheds a novel light on the history of
peoples such as Jews, Icelanders and the inhabitants of the British Isles. It
records the genetic impact of male dynasties like those of the Mongols and
the Manchus. And for those who know to ask the right questions, it retains
the secret family history of individuals such as Thomas Jefferson.

The compilers of the book of Genesis did their best, from available
myths and legends, to frame a coherent account of human origins. They
sought to address such questions as why people speak so many languages,
suffer pain in childbirth and wear clothes to conceal their nakedness. Human origins can now be explained in another way. Given that so little has
been preserved of the distant human past, it is remarkable how much is now
being retrieved. Many of the findings described here have been made in the
last few years. Though the frontiers of science are turbulent, throwing up
many claims that require revision in light of further evidence, the flood
of new findings described in these pages includes many unmistakable advances. The biological framework of human origins and nature is beginning
to emerge with surprising clarity as the record of past evolutionary change
now streams forth from the sequence of the human genome. In the long

search to understand ourselves, our obscure origins, our strange and contradictory nature, and the fragmentation of the once united human family into different races and warring cultures speaking thousands of different languages, we can begin at last to comprehend the long darkness before the dawn.

2

METAMORPHOSIS

In each great region of the world the living mammals are closely related to the extinct species of the same region. It is therefore probable that Africa was formerly inhabited by extinct apes closely allied to the gorilla and chimpanzee; and as these two species are now man's nearest allies, it is somewhat more probable that our early progenitors lived on the African continent than elsewhere.

CHARLES DARWIN, THE DESCENT OF MAN

FIFTY THOUSAND YEARS AGO, in the northeastern corner of Africa, a small and beleaguered group of people prepared to leave their homeland. The world then was still in the grip of the Pleistocene ice age. Much of Africa had been depopulated and the ancestral human population had recently dwindled to a mere 5,000.

Those departing, a group of perhaps just 150 people, planned to leave Africa altogether. Forsaking their familiar habitat was a serious risk since, as hunters and gatherers, their survival depended on intimate knowledge of local plants and animals. Nor is long distance travel ever easy for foragers who own no pack animals and must carry every necessity—weapons, infants, food and water.

The emigrants faced another danger in the world beyond. The lands outside Africa were not unoccupied. About 1.8 million years ago, during a warm interlude before the Pleistocene ice age began, early humans had left Africa in one or more migrations. Once separated from the main human population in the African homeland, these archaic people had followed their own evolutionary paths and in the course of time had become the distinct species known as *Homo erectus* and *Homo neanderthalensis. Erectus* settled in East Asia. The Neanderthals occupied Europe and intermittently parts of the Near East.

The Neanderthals in particular were formidable adversaries. They had large brains, larger in absolute size even than those of contemporary people, and were heavily muscled. They had developed serious weaponry, including stone-tipped thrusting spears. They surrounded or occupied the main exit point from Africa at the southeastern corner of the Mediterranean, including the area that is now Israel. The human lineages evolving in Africa may have tried many times to escape into the world beyond. But none had succeeded, and the Neanderthals' encirclement of the exits from northeast Africa seems a likely reason.

Why was the little group that left 50,000 years ago able to succeed when all earlier emigrants had failed? What drove them to take such a chance? What ties bound these people together and gave them the means to prevail?

To address such questions requires stepping back to the point in evolutionary time when the human line of descent split from the chimpanzee line. This is the moment at which our ancestors started to acquire the first of the adaptations that differentiated them from apes. These changes affected not just physical form but also the set of behaviors that together make up human nature. Of particular importance are the social behaviors, because both apes and people survive not as individuals but in social groups. In this sense the essence of human evolution is the transformation of ape society into human society.

Given the acute social intelligence of chimpanzees and bonobos, the emergence of the human society was perhaps not so great a leap. But the human lineage had the fortune to move down evolutionary paths that enlarged the brain and made possible the acquisition of language. The reason the ancestral human population was eventually able to burst out from its homeland seems to have been that 5 million years after having parted company with apes, it had at last perfected this critical component of human sociality.

Transforming Ape Society into Human Society

The ape society from which humans evolved lived some 5 million years ago somewhere in equatorial Africa. No fossil remains of these ancestral apes have yet been identified with certainty. Yet much can be inferred about them through the study of the two other living species descended from them, chimpanzees and their cousins the bonobos.

There are several reasons to suppose that the ancestral apes were very chimplike—an important assumption, if true, because it means that today's chimpanzees serve as a reasonably close surrogate for them. One is that gorillas, which split off from the ape lineage before the human-chimp split, are themselves rather chimplike, suggesting that so too was the lineage that led to the human-chimp ancestor. Another is that the earliest fossils on the human side of the chimp-human split are quite hard to distinguish from chimpanzees. A third reason is that the chimps of west and east Africa are very similar to each other, in both looks and behavior, despite having split apart 1.5 million years ago. Given that they have changed hardly at all in the last 1.5 million years, they may well have been just as conservative in the previous 3.5 million years.

The likely reason for this lack of change is that chimps still live mostly in forest, as did the joint ancestor, whereas the human lineage at some early stage left its forest home and took its chances in the open woodland, adapting to a quite different set of challenges. Chimps could stay much as they were because they were never under great evolutionary pressure to adapt to new environments.

If no fossils of the joint ancestor have yet been found, how can anyone know when it lived? The answer comes from genetics. By estimating the number of differences between corresponding stretches of DNA in the great apes and people, geneticists can construct a family tree whose branches are proportional in length to the evolutionary distances between the various species. The tree implies that the split between chimps and people occurred just over 5 million years ago (the most recent estimate suggests between 4.6 and 6.2 million years ago).[*3]

Genetic comparison also indicates why it is that chimps are the closest living species to humans. The chimp branches show four living subdivisions—and a fifth if the bonobo is counted—whereas the human branch is unnaturally straight, as if all competing human lines had fallen extinct, perhaps because they were pruned away by members of the one surviving lineage.[4]

The estimate of 5 million years for the chimp-human divergence fits quite well with a salient event in the earth's climatic history. Global climate

---

*A genetic tree can be anchored in time if any one of its branch points is datable from the fossil record. In this case the divergence of orangutans from the other species is known from fossils to have occurred some 12 to 16 million years ago. This gives a date for the human-chimp branch point. Gorillas split off 7.3 million years ago, some 2 million years before the chimp-human divergence.

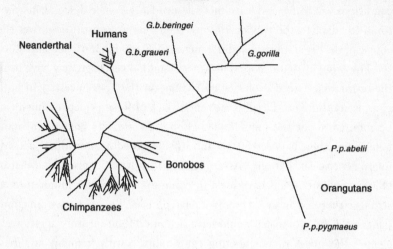

FIGURE 2.1. FAMILY TREE OF HUMANS AND OTHER GREAT APES.

The figure shows the genealogy of an extended family—that of orangutans, gorillas, chimpanzees and humans. The tree was constructed by decoding part of the mitochondrial DNA of each species and comparing the common sequences of DNA letters, looking for differences. The lengths of the tree's branches are proportional to the number of differences that have evolved between each species.

The bushiness of the chimpanzee branch reflects the amount of genetic diversity that has evolved in the chimp line. By comparison, the human branch is unnaturally straight, a sign of natural calamities that reduced the human population, or of modern humans having killed off all rival human species, or both. The Neanderthal sequence is derived from DNA extracted from fossil bones.

cooled between 10 and 5 million years ago, with the period from 6.5 to 5 million years being particularly harsh. Water was locked up in massive glaciers, and sea level fell so low that the Mediterranean sea was repeatedly drained, depriving Africa of a source of moisture. In the dry, cold climate, the equatorial forests shrank and in places fragmented into woodland.[5] In such a habitat, with open canopy and large spaces between trees, forest tree dwellers would have had to spend more time on the ground, at considerably greater risk from large predators. These cruel years placed such stress on the forest apes that many went extinct.

It is just this kind of environmental stress that has forced evolutionary change since the beginning of life on earth. In response to new conditions, a species must adapt, by drawing on the genetic variation available in the members of its population. The individuals better endowed to meet the new conditions will thrive and leave more descendants. As the descendants continue to adapt to the new conditions, their genetic constitution, over the course of the generations, will differ increasingly from that of their ancestors.

The drought that occurred in Africa 5 million years ago may have been the agent that forced evolutionary change on the apes ancestral to both chimps and humans. Despite the serious lack of fossil evidence, much can be inferred about the joint human-chimp ancestor. Its population numbered somewhere between 50,000 and 100,000 breeding individuals, according to genetic calculations.[6] Assuming that its way of life resembled that of chimpanzees, it would have lived in communities about a hundred or so strong, structured around a group of related males. By inference from chimpanzee society, the males would have defended their territory aggressively against the males of neighboring communities, with frequent fatalities. Each community's survival strategy lay in defending as large as possible an area of fruit trees for its females to feed in.

The male apes, again on the assumption the species resembled chimpanzees, were much larger than the females and paid little attention to them except to mate. There was no particular bond between male and female. Each sex had its own hierarchy, with the females subservient to the males. The males spent their day, when they weren't fighting, building alliances with other males and trying to work their way up the male hierarchy of their community. The risks of being alpha male were considerable, but the payoff considerable, at least in Darwinian terms: the alpha male and his allies got to father most of the community's offspring.

So consider this population of 100,000 chimplike apes somewhere in the eastern side of equatorial Africa 5 million years ago. Times are tough and their forest homeland is shrinking. The trees no longer carry enough fruit. The apes are forced to spend a lot of time on the ground searching for other sources of food. Large cats stalk or ambush the unwary. Each generation is tested by this harsh new environment, and in each generation the better adapted produce more offspring.

There are two kinds of survivor. One, clinging to the remnants of forest, manages to continue in much the same way of life: this is the lineage that leads to chimpanzees, and because it clings to the same habitat it has no great need to change its way of life or physical form.

The other manages to survive by venturing into a new niche—it learns to occupy both the trees and the new spaces that have appeared in between them. Helping it survive on the ground is the emergence of a critical new ability—that of walking on two feet.

Two-footedness—bipedalism in paleoanthropologists' parlance—is the first great stride toward becoming human. Standing upright is in fact not so big an adjustment for apes, who move in the trees by hanging under branches and swinging from one arm to the other. Monkeys, on the other hand, prefer to run along the tops of branches; so when some of them started to live on the ground—becoming baboons—they preserved their four-footed style of travel.

Chimpanzees get around on the ground by knuckle-walking—using the knuckles of the hand as front feet. So why did those on the human side of the split prefer bipedalism? Many advantages of bipedalism have been cited as decisive—it frees up the hands for carrying things, it allows better surveillance of the surroundings—but the most likely reason for its emergence is simply that upright walking is more efficient than knuckle-walking. For the same expense of energy, a chimp can knuckle-walk 6 miles a day but a man can walk 11.[7] Bipedalism probably evolved because it was a better way of getting about, and its other benefits were at first incidental.

The first walking apes, woodland primates known as australopithecines, appear in the fossil record 4.4 million years ago. A breathtaking trace of their presence is a line of footprints made nearly a million years later at Laetoli in Tanzania. The tracks of two individuals, perhaps a parent and child, extend for 165 feet across the ash from a nearby volcano, crossed by the tracks of other animals, perhaps fleeing from the eruption. A few frozen seconds of time, with prints that look so human.

Yet, walking and feet aside, the australopithecines seem to have been mostly apelike. With long arms, they retained the ability to move in trees. Their brains were only slightly larger than an ape's. And as with apes, the sexes were of very different sizes, the males being much larger.

Larger male size, in primate societies, reflects competition between males for females, and is particularly prominent in gorillas, whose harem-keeping males are twice the size of females. Male chimpanzees are 25% larger than females, but in today's human populations men are only 15% larger than women. Male australopithecines were about 50% larger than females, suggesting that australopithecine society was much like that of chimpanzees, with strong rivalry between males and a separate male and female hierarchy. For two million years of australopithecine existence, there is little sign of human form, apart from the critical upright gait, and no reason to assume that social behavior had changed much from the chimpanzee-like pattern.

Then, from 3 to 2 million years ago, there was another long period of cool, dry climate in which Africa's forests shrank once more, and many species adapted to living in them fell extinct. The changing climate also put pressure on the australopithecines to develop new sources of food. Their diet, to judge by the nature of the microscopic wear on their teeth, was mostly vegetarian until 2.5 million years ago. At this time the australopithecines, already adapted to living in open woodland, had evolved two quite separate solutions to the problem of survival, according to the evidence of their fossil remains. One of the two new species, known as the robust australopithecines, had developed larger cheek teeth, suitable for eating coarse leaves. The other had emerged with a much more original solution than chewing away at vegetation. It seems to have decided to try its hand at carnivory. Meat-eating allowed for a smaller gut and furnished the extra nutrition that made possible a larger brain.

This second species is known as *Homo habilis*. The title of *Homo* is one it does not clearly deserve since, far from being fully human, it retained its apelike body form and still used the trees as a refuge. But it possessed a striking new adaptation. The australopithecines had lived for 2.5 million years with brains scarcely bigger than a chimp's, but with *habilis* the brain at last started to expand. Chimpanzees' brains have a volume of 400 cubic centimeters, compared with the 1,400 ccs of the average modern human brain.[8] The australopithecine brain size ranged from 400 to 500 ccs. The brain volume of the known *habilis* skulls ranges from 600 to nearly 800 ccs.*[9]

For a species to put resources into growing extra neurons is not as obvious an investment as it may seem. Brawn and teeth count a lot in the strug-

---

*Brain volume is a crude measure because it reflects many other things besides raw computing power, including body size and living in a cold climate. The Inuit, or Eskimos, have the largest brains of any modern humans, and Neanderthals had the largest brains of all. Paleoanthropologists often prefer to use a different measure, called the EQ, or encephalization quotient, which gives a measure of brain volume in relation to body volume. By this measure the numbers are as follows: chimpanzees 2, australopithecines 2.5, *Homo habilis* 3.1, *Homo ergaster* 3.3, modern humans 5.8. Source is ref. 9.

gle for survival. Brain cells are greedy consumers of glucose and oxygen. The modern human brain is only 3 percent of the body's weight but uses some 20% of the energy required for metabolic maintenance. "When costs are taken into account, the rarity of the human evolutionary phenomenon is at last understandable," writes the anthropologist Robert Foley.[10]

It's easier to explain how *habilis* sustained its larger brain than why it got it. Brains require a high quality diet to sustain them, such as meat but not vegetation can provide. Meat-eating requires less tooth power than does chomping through mounds of vegetation and *habilis* indeed had smaller teeth. And *habilis* appears on the scene at the same time, 2.5 million years ago, as do the first stone tools. If, as seems likely, *habilis* was the maker and user of these implements, that would explain its smaller teeth and how it managed to nourish a larger brain; it didn't need large teeth because it was using tools to hunt or scavenge meat, and the richer diet supplied the energy for its greater cognitive capacity.

Still, that doesn't explain what specific environmental forces made a larger brain advantageous in the first place. Higher social primates like apes and people probably encounter no problems more challenging than those of dealing with other members of their community. If so, the likeliest reason for *habilis*'s greater brain size would have been increasing social complexity.

The stone tools associated with *habilis* are known, rather grandly, as the Olduwan Industrial Complex since they were first found in the Olduvai Gorge in eastern Africa. The tools consist mostly of pebble cores and the rough flakes struck off them. They have a kind of random appearance, as if the maker was not holding any design in mind and was content to accept whatever shape of stone nature might produce. Still, these random pieces of rock would have been useful for a wide variety of purposes, such as cutting through hide and ripping the flesh off of bones.

The technology of the Olduwan Industrial Complex seems to have represented the limit of *Homo habilis*'s new cognitive capacity and inventive powers. Far from being followed by further innovations, it remained unchanged for 800,000 years. This lack of development in stone tool-making may reflect a similar conservatism in the lifestyle of its maker.

The emergence of bipedalism and the beginnings of a larger brain were two major genetic steps in the process of morphing the chimplike ancestor into modern people. A third genetic revolution occurred 1.7 million years

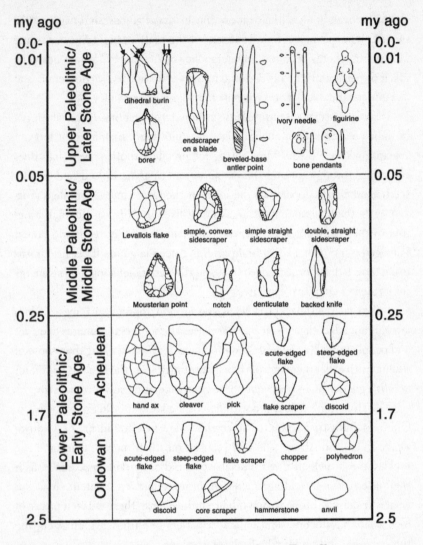

FIGURE 2.2. EVOLUTION OF HUMAN STONE TOOL KITS.

The first stone tools, made by *Homo habilis*, appeared 2.5 million years ago. The kit remained in use, unchanged, until 1.7 million years ago, when it was replaced by a more sophisticated set of implements, the Acheulean industry, made by a more advanced species, *Homo ergaster*.

The time axis, expressed in units of millions of years ago, is not to scale. The lower two bands occupy a time span of 2.25 million years, the upper two bands one of just 0.25 million years. During this latter period, the conservatism of the previous 2 million years was replaced by a much brisker tempo of innovation. The Acheulean gave way to the Middle Stone Age tool kit, made by both the Neanderthals in Europe and the human lineage in Africa. Then from 50,000 years ago, the modern humans who replaced the Neanderthals in Europe started making the highly refined artifacts of the Upper Paleolithic age. These included smaller tools, some designed to be set in wood handles or weapons, as well as decorative and artistic objects.

ago in the form of the physical and behavioral changes shown by a new species known as *Homo ergaster*. *Ergaster* was presumably a descendant of *habilis*, though the fossil record is too scant for proof. It's the first creature whose skeleton shows most of the features of human identity even though its brain volume, at 800 ccs, is way below the modern capacity.

Ergaster's arms were of human length, not ape length, suggesting it had made a final farewell to the trees and was committed to living at ground level. Its chest cavity had the human barrel shape, not the cone shape of an ape's, the indication of a major change in diet. Apes need enormous guts to digest masses of plant material and their rib cages are cone-shaped because they must be wide enough at the bottom to cover the stomach compartment. The barrel shape of *ergaster*'s chest was positioned above a smaller belly, showing that it was eating a richer diet, consisting of meat and maybe also tubers, the starchy roots that served as storage devices for plants living in dry environments.[11]

Tubers, a staple of hunter-gatherers, are a likely new food because *ergaster* had learned for the first time to inhabit dry, hot areas in East Africa where tubers flourish. Aridity, shown by the presence of dust in sea-floor sediments, increased sharply at the time of *ergaster*'s emergence. *Ergaster* may even have learned to cook the tubers, a significant advance if so because cooking releases nutrients from foods, making them more digestible.[12] There is no evidence that *ergaster* did in fact use fire, but since ashes do not preserve well, the evidence of *ergaster*'s cookouts may have been lost.

Paleoanthropologists see signs in *ergaster*'s body structure that it had made a significant transition from an apelike to a more human mode of social organization. *Ergaster* is the first species along the human lineage to show a sharp reduction in male size compared with female, although the females are still smaller. This is a hint of some important change in social structure, very possibly a switch from the separate male and female hierarchies of chimp communities to the male-female bond that characterizes human societies.

Such a shift does not imply anything as extreme as monogamy, but it could mark at least the beginning of a family structure in which males took some interest in protecting and feeding the mothers of their children. Tubers, assuming they had become part of *ergaster*'s diet, are items that women can collect. A greater role for women in providing food might have promoted the pattern of greater cooperation between the sexes, as is implied in the decreased male-to-female size ratio.

Further evidence of change in relations between the sexes comes from *ergaster*'s anatomy. Its pelvis was narrower than *habilis*'s, and the smaller birth canal meant that much of the infants' extra brain size had to be acquired by growth after birth. This in turn means that infants would have had to be carried, making the women more vulnerable. The fathers might have found that in order to protect their genetic legacy they needed to spend more time guarding their specific offspring and the children's mother, not just the group's general territory. Such behavior, and a closer bond between parents, would have been genetically favored if it led to more surviving offspring. That's a long chain of inference from the simple anatomical fact of a narrower pelvis, but not so unreasonable.[13]

Ergaster's brain was only slightly larger, in relation to body size, than that of *habilis*, its assumed predecessor, but it was nonetheless capable of a whole new level of stone tool-making. The more sophisticated tool kit, called the Acheulean Industrial Tradition, included the lozenge-shaped stones thought to be hand axes, although their precise use is not known for sure, as well as cleavers and other large tools. From the microscopic pieces of material on the stones, as well as modern experiments in stone knapping, the tools seem to have been designed for a wide array of tasks, such as heavy and light duty butchering, slitting hides, breaking bones, cutting grass, and woodworking. *Ergaster* doubtless used many tools and materials made of wood and other perishable substances.

The Farewell to Fur

The humanlike species that evolved during the first three million years after the split with chimps probably looked far more apelike than human. Their bodies were of ape proportions, with long arms, and doubtless covered with hair from head to toe. Not until *ergaster*'s arrival on the scene did the lineage's physical appearance assume a more recognizably human form. *Ergaster*, unlike its predecessors, possessed an external nose. This acquisition was its most prominent adaptation to hot, dry climates, the role of a nose being to conserve water by cooling and condensing moist air from the lungs before it leaves the body.

After the differences in body proportions, humans differ most strikingly from their ape cousins in the distribution of body hair. Over most of the body people are essentially hairless, possessing just vestigial hair that is largely invisible. Nakedness is a complex issue, for which there may be several layers of explanation. Hairiness is the default state of all mammals, and the handful of species that have lost their hair have done so for a variety of compelling reasons, such as living in water, as do hippopotamuses, whales and walruses, or residing in hot underground tunnels, as does the naked mole rat. With humans, the prime cause may have been the need to sweat. *Ergaster* may have been the first of the human line to shed its fur in favor of naked skin, in the view of the paleoanthropologist Richard Klein.[14] His inference is based on the idea that if *ergaster* were living in dry, hot places, it would need to have evolved a way of cooling the body and its larger brain. Sweating, an efficient way to do this, requires a naked skin. Besides, humans must have lost their hair at some time, and the most plausible period is when they traded the shade of the trees for the heat of the savanna.

Another, perhaps secondary, reason for human nakedness may have to do with sexual preferences. Darwin, who first suggested the idea, gave the matter serious attention in his book of 1871, *The Descent of Man.* "May we then infer that man became divested of hair from having aboriginally inhabited some tropical land?" Darwin asked. (He had already assumed that humans originated in Africa, because that is where the great apes are found, but there was then no fossil evidence to confirm the idea.) Yet that couldn't be the whole story because other primates in tropical countries have retained their hair. Perhaps shedding the body hair freed humans from the burden of parasites like lice, fleas and ticks. But that didn't seem a decisive enough advantage to Darwin. "The view which seems to me the most probable," he concluded, "is that man, or rather primarily woman, became divested of hair for ornamental purposes." [15]

Darwin believed that sexual selection was an important factor in evolution because it determined mating success. Sexual selection arises in two distinct forms, intersexual and intrasexual. The first is the way that men and women choose each other as mates; the second is the competition within each sex, between men for women and, sometimes more discreetly, between women for men. Hairlessness would have been favored, in Darwin's view, if men and women had preferred partners with less hair. Two biologists, Mark Pagel and Walter Bodmer, have recently reinvoked Darwin's idea of sexual selection as the driver of human hairlessness. They suggest that lack of hair was favored among early humans because it was a sure signal that no parasites were lurking in their fur.[16]

The date proposed by archaeologists for when humans lost their hair is based on the guess that it coincided with the emergence of *Homo ergaster*. But an actual fix on the date has been supplied by geneticists. Their research illustrates the wealth of information that can be extracted from a single gene if the right questions are asked.

The gene in question is one that makes the melanocortin receptor, a protein that helps determine skin color. It does so by controlling the proportions of different-colored melanin pigments that are synthesized in a person's skin cells. Some versions of the melanocortin receptor produce black skin and hair, others generate ginger or brown or yellow.

Rosalind Harding, of the University of Oxford in England, recently analyzed the order of the DNA units in the melanocortin receptor gene possessed by people from Africa, Europe and Asia. She and her colleagues found that all Africans had essentially the same version of the gene but that people outside of Africa possessed many different versions.[17]

An obvious explanation for the receptor gene's constancy in Africa is that it is under fierce selective constraint there, meaning that natural selection prevents any significant change. The African version of the gene is set to produce maximum blackness; any change in its DNA sequence is likely to make the skin lighter and its owner more vulnerable to the sun's ultraviolet radiation, which destroys an essential nutrient known as folic acid. (Ultraviolet radiation can also cause skin cancer, but it is the destruction of folic acid that is more likely to reduce fertility and hence to shape the evolution of the gene.) Anyone with a changed melanocortin receptor gene is likely to leave fewer or no descendants, and the variant gene will in time be eliminated from the population. Hence everyone living under the African sun has the same version of the gene, no deviations allowed.

Before human ancestors lost their hair, however, their skin was almost certainly pale, according to Nina Jablonski, an expert on the evolution of human skin color.[18] This can be inferred from the skin color of chimpanzees, the reliable surrogate for the joint human-chimp ancestor. Beneath their dark hair, which protects them from the sun, chimpanzees have light skin. They too have a melanocortin receptor gene, but it exists in many different versions, as if natural selection does not mind letting it vary, and all produce pale skin. (Chimps have dark-skinned faces, but that is from tanning of the pale faces they have at birth.)

Reading Harding's article, Alan Rogers, a population geneticist at the University of Utah, wondered how African populations had all acquired the same version of the gene. The process must have started, he supposed, when the human lineage first started to lose its apelike hair, dangerously exposing the pale skin beneath. Any mutation in the melanocortin receptor gene that led to a blacker, more protective skin would have conferred a great advantage on its owner. In several generations the new version of the gene would sweep through the population.

Genetic sweeps can often be dated because after a must-have gene has become universal it starts to accumulate what are known as silent mutations, ones that don't alter the structure of the gene's protein and so are not eliminated through natural selection.* Since the silent mutations accumulate at a known rate, the number of them is a measure of the time that has elapsed since the new version of the gene swept through a population.

Rogers realized that from the silent mutations in the African version of the melanocortin receptor gene, he could calculate the date at least of the gene's most recent sweep. He estimates that this event took place about 1.2 million years ago.[19]

There may have been several earlier such genetic sweeps, each one producing a progressively more effective version of the melanocortin receptor gene. The gene, after all, had to make a very significant transition, from producing the pale skin of the joint human-chimp ancestor to the black skin that protected the newly hairless body from the sun in the scantly shaded African savanna. If the first of these sweeps had started several thousand

---

*Genes are strings of DNA that embody the information to make proteins, and proteins are the working parts of the living cell. But only some of the DNA in a gene contains the code for its protein; the rest is known as noncoding DNA. A mutation in the coding DNA usually interferes with the protein's structure. But mutations in the noncoding DNA usually have no effect on protein structure and are called silent.

years before, that would fit well with the archaeological evidence for the
emergence of *Homo ergaster* 1.7 million years ago.

Outside of Africa, to jump a little ahead in the human story, the
melanocortin receptor gene became free once more to collect mutations and
become less efficient at triggering the black, radiation-protective form of
melanin. That could well have been an advantage for people living in cold
northern climates since they require extra exposure to the sunlight that is
needed to help synthesize sufficient quantities of vitamin D, the lack of
which causes misshapen bones and the disease known as rickets. In every
population of the world, women's skin color is 3 to 4% lighter than men's,
perhaps through sexual selection by men, and perhaps because of mothers'
greater needs for vitamin D.[20]

To turn to another curious feature of human hair, when did you last see
a chimpanzee getting a haircut? Human head hair differs from that of apes
in that it never stops growing. If the hair follicles on the human head behaved like those of chimpanzees, they would follow an orderly cycle in which
each would grow a hair for several weeks; the hair, after reaching a certain
length, would then be shed, and the follicle would grow another hair. With
people, this cycle has been lengthened from weeks to years.

The reason that uncontrolled hair growth was favored by natural selection may have been that it offered a means of signaling copious amounts of
social information. In every society in the world, people spend an inordinate
amount of time in cutting, shaping, braiding, plaiting, curling, straighten-
ing, decorating and otherwise gussying up the appearance of their hair.
Much the same is true of men's beards and mustaches. To let one's hair grow
unkempt is a sign of the outcast, or that one is in deep mourning. Trimmed
hair sends a variety of important signals about the wearer's health, wealth
and social status. But for all this social signaling activity to occur, humans
had first to abandon the self-maintaining hairdos of other apes and acquire
hair that required continual attention.[21]

Geneticists have calculated a date for the birth of the hairdressing industry. Keratin, the protein in hair, comes in a large number of different varieties, each prescribed by a different gene. Humans, chimpanzees and
gorillas have much the same set of keratin genes but with a striking difference. One of these genes, the human version of which is known as phi-
hHaA, produces a working keratin protein in chimpanzees and gorillas but is

inactive in people of all ethnic groups. Although the genetic regulation of human hair growth is not yet understood, it seems likely that the inactivation of the phi-hHaA gene is the step by which the hair follicles of the human head have escaped from the orderly cycle imposed on chimp and gorilla hair follicles. By comparing the mutations in various versions of the human and ape gene, researchers have calculated that the human version became inactivated some 200,000 years ago.[22] This is long after *Homo ergaster* had become extinct and about the time that the human line acquired its contemporary physical appearance.

The First Exit from Africa

Because *ergaster* was adapted to living in dry places, it could survive in many environments. This adaptability made possible a momentous step, the first spread of human lineages to the lands beyond Africa. A close relative and presumably descendant of *ergaster*, known as *Homo erectus*, had reached Asia by at least 1 million years ago and maybe much earlier—stone tools recently found in northern China have been dated to 1.66 million years ago.[23] By 1 million years ago, *ergaster* itself had reached both the northern and southern extremities of Africa. And by at least 500,000 years ago a human lineage had reached Europe, perhaps through a second migration from Africa of another *ergaster* descendant known as *Homo heidelbergensis*. In Europe, under the glacial conditions that prevailed from 400,000 to 300,000 years ago, these new migrants evolved into *Homo neanderthalensis*, the Neanderthals, broadboned, thickset people who were adapted to the cold.

Erectus and the Neanderthals are referred to as archaic humans in distinction to the human lineage that remained in Africa and ultimately became modern. With the departure of the archaics from Africa, the human gene pool was split into three main branches, in Africa, Asia and Europe, and each now followed a separate evolutionary path.

In Africa, it was not until 500,000 years ago, more than a million years after *ergaster*'s first appearance on the scene, that brain size relative to body size increased significantly, and not until 200,000 years ago that it reached the contemporary standard.

Yet a curious feature of the steadily increasing brain size of the human

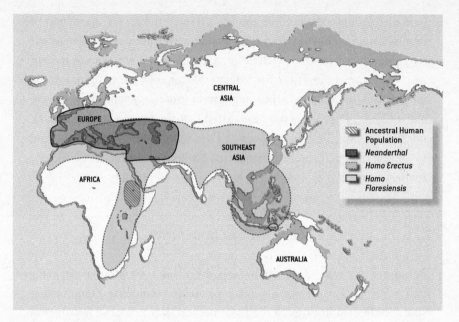

FIGURE 2.3. THE THREE HUMAN SPECIES OF 50,000 YEARS AGO.

The world 50,000 years ago was occupied by three human species—*Homo erectus* in East Asia, the Neanderthals in Europe, and the ancestral human population in northeast Africa. In addition *Homo floresiensis*, thought at present to be a downsized version of *Homo erectus*, lived on the island of Flores, in Indonesia.

Because of the ice age conditions that then prevailed, sea level was some 200 feet lower than at present and land area was larger, as shown by the shaded areas round the continents.

The range shown for *Homo erectus* encompasses sites that range from 1.7 million to 50,000 years ago in age. The species probably did not occupy all of this range throughout the period, and toward the end of it was probably found mostly in southeast Asia. The range and location indicated for the ancestral human population is conjectural.

lineage is that it was not accompanied by any significant change in behavior that is visible in the archaeological record. Just as the Olduwan stone tool kit remained unchanged from 2.5 million to 1.7 million years ago, the Acheulean tool kit that succeeded it was also almost unvaried from its emergence 1.7 million years ago until its disappearance about 250,000 years ago. *Erectus* in Asia may have varied the formula by using bamboo, which is hard but perishable, in place of stone. This would explain the strange absence of Acheulean hand axes in the Far East. The Neanderthals in Europe used the same tool kit as the human lineage in Africa. "The technologies of these archaic creatures were homogeneous across and even between whole continents. The Acheulean, for example, although varying in minor form, is known from Cape Town to Cardiff," writes the evolutionary anthropologist Robert Foley.[24]

The Acheulean stone tool kit was followed by one that archaeologists use to define the Middle Stone Age in Africa and the Middle Paleolithic or Mousterian in Europe. The makers of the Middle Stone Age tools were the descendants of *ergaster*, on the way to becoming large-brained *Homo sapiens*, while Mousterian artifacts are the handiwork of *ergaster*'s European cousins, the Neanderthals. The tool kits in both continents are very similar, and both differ very little from the Acheulean. The principal difference is the absence of the characteristic Acheulean hand axes. Perhaps archaic humans learned how to mount smaller stones on handles, and these composite wood-and-stone tools replaced the hand axes.[25]

What stopped the Middle Stone Age people from leaving Africa as their predecessors had done? It seems the descendants of these predecessors may have been hemming them in. Even by 100,000 years ago, the human lineage in Africa was still using the same tool kit as the Neanderthals in Europe and as *Homo erectus* in Asia. They evidently enjoyed no competitive advantage over the Neanderthals.

During a warmer period in the ice age that lasted from 125,000 to 90,000 years ago, people came close to escaping from Africa. They extended their range to the region that is now Israel, at the border of Africa and Asia. But during the cold period that prevailed from 80,000 to 70,000 years ago, the Neanderthals expanded their range southward to western Asia and seem to have destroyed the emigrants.[26]

The humans who lived during the African Middle Stone Age, which lasted from 250,000 to 50,000 years ago, had a way of life that was more sophisticated than their *ergaster* forebears but only slightly so. They obtained their stone locally, not through trade, suggesting they had small home ranges or very simple social networks. They hardly ever made things of bone, ivory or shell. They were not very good hunters and couldn't even fish. Their populations were small, as judged by the archaeologists' standard people-meter, the tortoise test. (People catch large tortoises first, then smaller ones. Tortoises are so slow to replace themselves that the size of a human settlement can be judged by whether the tortoise bones are large, indicating a sparse human population, or small, meaning rather more mouths to feed.)[27]

Like the Neanderthals, the Middle Stone Age people seem to have buried their dead, but very simply, and to have collected pigment making minerals, though for an unknown purpose. They left no clear evidence for art or decoration.

This pattern of behavior altered scarcely at all as one millennium followed another. Strangely, the human form was changing much more. In Africa, people began to attain the skull size and skeleton of contemporary humans some 200,000 ago. The oldest known specimens, from a site near the Kibish river in southern Ethiopia, may be about 195,000 years old,[28] and fossil remains of people with this new form start to be commonly found about 100,000 years ago.

Anatomically and Behaviorally Modern Humans

Modern human behavior, at least as judged by archaeologists, means behaving like living hunter-gatherers. By this criterion, the humans of 100,000 years ago did not behave like modern humans, even though they looked like them. They are known as anatomically modern humans to denote that they were not so in behavior.

What kept them from attaining a fuller modernity? The question of behavioral modernity is of great significance because it appears to be the last major step in the emergence of the ancestral human population. The components of modern behavior appear most prominently around 45,000 years ago in Europe. At sites throughout Europe, the staid culture of the Neanderthals begins to yield to a set of new and more inventive techniques. There is a new set of stone tools, more carefully crafted to attain specific shapes. There are complex tools made of bone, antler and ivory. The bringers of the new culture made personal ornaments, of materials such as punctured teeth, shells and ivory beads. They played bird-bone flutes. Their missile technology was much improved. They were avid hunters who could take down large and dangerous game. They buried their dead with rituals. They could support denser populations. They developed trade networks through which they obtained distant materials.[29]

This new modern culture is called the Upper Paleolithic. Some archaeologists have proposed that it was created by Neanderthals or by Neanderthals interbreeding with modern humans. It now seems more likely that the culture was the work of behaviorally modern humans alone, who simply replaced the Neanderthals, over a period of several thousand years, throughout their European domain.

One reason for this interpretation is that several diagnostics of modern behavior can be seen to have appeared first in Africa, in the Later Stone Age, which had begun by at least 46,000 years ago. (The Later Stone Age of Africa and the Upper Paleolithic of Europe are the same archaeological period but for historical reasons have different names in the two continents.) The timing suggests that humans with modern behavior first evolved in Africa and later reached Europe. This hypothesis, made on purely archaeological grounds, has been confirmed by the genetics of modern human populations, all of which point to a diaspora in recent times from an African homeland.

So if behaviorally modern humans arose in Africa, the final stage of human evolution in Africa was that from anatomically modern humans of 100,000 years ago to the behaviorally modern people who appeared some 50,000 years later. What caused that profound transition?

Archaeologists tend to explain changes in terms of culture. But paleoanthropologists, looking at much longer sweeps of time, are more accustomed to seeing evolution and genetic change as the principal shaper of novelty. The paleoanthropologist Richard Klein has proposed that the transition to modern behavior was so profound that it required a genetic change: "Initially, the behavioral capabilities of early modern or near-modern Africans differed little from those of the Neanderthals, but eventually, perhaps because of a neurological change, they developed a capacity for culture that gave them a clear adaptive advantage over the Neanderthals and all other non-modern people."[30]

It was obviously a genetic change, not a cultural one, that endowed the australopithecines with upright stature 4.4 million years ago. It was a suite of genetic changes 2.5 million years ago that remodeled the australopithecines into *Homo habilis* with its larger brain and tool-making ability. A third far-reaching genetic makeover 1.7 million years ago reshaped *habilis* into the more humanlike *erectus* and caused a behavioral transition from male and female hierarchies to the pair bond system. And it must have required a fourth genetic revolution, Klein believes, to make possible the emergence of behaviorally modern humans 50,000 years ago.

That genetic revolution was evidently profound enough to affect many different aspects of human social behavior and technical skills, all characterized by a striking new capacity for innovation. The most likely cause of such a transformation, in Klein's view, would have been the emergence of language.

For a social species, nothing could make a greater difference than the ability to transmit precise thoughts from the mind of one individual to another. Language would have made small groups more cohesive, enabled long range planning and fostered the transmission of local knowledge and learned skills.

It is certain that modern humans could speak before they left Africa, so language must have evolved sometime before 50,000 years ago, and after 5 million years ago when the human line split from that of chimpanzees. Looking in the archaeological record for some sharp increase in behavioral complexity that might signal the evolution of language, there are few likelier moments than the transition from anatomically modern to behaviorally modern humans.

Klein's argument is not universally accepted by other archaeologists, some of whom have attacked a principal element in his case, the sharp discontinuity he sees between the behaviors present at the end of the Middle Stone Age and the beginning of the Later Stone Age. Two critics, Sally McBrearty of the University of Connecticut and Alison S. Brooks of George Washington University, argue that there was a gradual accumulation of advanced behaviors throughout the Middle Stone Age that eventually added up to modern behavior. "As a whole the African archaeological record shows that the transition to fully modern behavior was not the result of a biological or cultural revolution, but the fitful expansion of a shared body of knowledge, and the application of novel solutions on an 'as needed' basis," the two archaeologists write.[31]

Klein's view that there was no modern behavior in Africa prior to 50,000 years ago, once uncontradicted by any evidence, has been challenged by several individual finds. Christopher Henshilwood of the University of Bergen in Norway recently found a set of 41 shells, all perforated in the same way as if meant to serve as beads on a necklace. The shells were excavated from the Blombos cave in southern Africa; the sand in which they were found has been dated to 76,000 years ago by a physical technique.[32] Dates of this age are beyond the reach of the reliable radiocarbon method, and the methods used instead are considerably less accurate. If the 76,000 year date is right, however, and if the pierced shells are indeed beads, it would mean that decorative art, a practice associated with behaviorally modern humans, began much earlier than supposed.

Another behavior generally considered modern is fishing. Hence a possible problem for the Klein position is eight barbed points made of bone, which could have been used to harpoon fish. The points come from the Katanda riverside site in Zaire from strata about 100,000 years old. Klein believes that the bone points, if directly dated, would turn out to be less than 12,000 years old.

A difficulty for Klein's case has long been that of the very early date by which modern humans apparently reached Australia. Since this feat required boat building and some navigation skills, it would certainly count as a modern behavior. The date of human arrival in the continent has long been set at 60,000 years ago, based on a burial at a site near Lake Mungo in southeastern Australia. The finding indicated that modern behavior had been attained in Africa even earlier. But the Lake Mungo date recently turned out to be incorrect: the burial site is only 42,000 years old, with artifacts suggesting an earlier human presence at somewhere between 50,000 and 46,000 years ago.[33] This date fits quite well with the theory that behaviorally modern humans were able to leave Africa only 50,000 years ago.

As for the barbed points and other artifacts, Klein argues that they are, at least for the moment, anomalies that don't fit into the established archaeological pattern. If the Katanda points are 100,000 years old, why didn't such an important technique as fishing spread like wildfire? Yet no other African site shows evidence of fishing until 25,000 years ago. Klein, who does his fieldwork in Africa, has twice found sheep bones in strata belonging to the Middle Stone Age, which ended 50,000 years ago. Since sheep were not domesticated for another 40,000 years, the bones are clearly intrusions from a higher level, introduced by burrowing animals or one of the many other sources of confusion in the archeological record. Archaeologists must expect to find a few later intrusions in any stratum, Klein believes, and should therefore base their conclusions on well established patterns, not on the occasional anomaly. His critics, he believes, are looking at the noise in the record, not its true signal.

Looking at the extraordinary process by which apes were slowly molded into humans, it is easy to think of the end result as some goal that evolution was driving toward. But evolution, of course, is a blind, inanimate process with no goals, let alone any interest in human welfare. It is driven by mutation, natural selection and drift. Mutation—random natural changes in the chemical units of DNA—is the ceaseless generator of novelty in the human genome. That novelty is the raw material on which natural selection acts, rejecting changes for the worse and retaining those that confer reproductive advantage. The mighty tide of genetic drift, through the random selection of genes between generations, makes some genetic variants a permanent fixture in a population and extinguishes many others, reducing the novelties that mutation introduces.

The interplay of these three forces may sound like a recipe for chaos, yet evolution's mechanisms do in fact bring into being, over the course of long periods of time, structures of extraordinary complexity, such as the human ear or eye. Because such adaptations are ones that human engineers could create only by design, biologists often talk about evolution as if it possessed intent or forethought. But this is just a shorthand way of referring to the evolutionary process and is not meant to imply that evolution has any goal in mind.

In the sense of the biologists' shorthand, it could be said that with the development of language, evolution had accomplished a major part of the task of morphing an ape into a human, and of shaping humans into a truly social species. Since language is such a defining faculty of modern humans, providing perhaps the only clear distinction between people and other species, its nature and evolution merit a closer look.

3

FIRST WORDS

[Language] certainly is not a true instinct, for every language has to be learnt. It differs, however, widely from all ordinary arts, for man has an instinctive tendency to speak, as we see in the babble of our young children; whilst no child has an instinctive tendency to brew, bake or write.

CHARLES DARWIN, *THE DESCENT OF MAN*

EVOLUTION'S RAW MATERIAL is the gene pool of a species and the mutations that arise at random in those genes. This formidable constraint means that an organ or faculty cannot be created out of nothing; it can only be shaped, by gradual stages, out of some existing structure, and each of those intermediate stages must confer advantage in its own right.

One reason why human language is so deeply puzzling to biologists is that it seems to defy this rule. It is a vibrant, fully developed faculty in people, but is not possessed, even in rudimentary form, by any other species. It seems to have popped up into the recent human line from nowhere.

The origins of language would perhaps seem somewhat less mysterious if our archaic cousins, the Neanderthals and *Homo erectus*, had survived to tell what kind of communication skills they commanded. But these branches of the hominid tree have been docked, leaving only one survivor.

Primatologists have therefore looked for the roots of the language faculty in social primates such as apes and monkeys. These species do indeed possess many of the neural systems that are needed in support of language. They can make a wide range of elaborate sounds. They have acute senses of hearing with which to perceive and analyze the sounds made by members of their own species. As for thought, there is no doubt that the social primates are capable of quite elaborate cognitive processes, such as those required in keeping tally of who one's relations are, who owes one favors, and where one stands in the social hierarchy.

But despite possessing much of the neural equipment for speech, monkeys and apes simply lack the ability to translate their thought into anything resembling human language.

Several primate species have communication systems of considerable sophistication. Gelada baboons have 22 different kinds of call, and gorillas have been recorded using some 30 different gestures.[34] One of the best studied animal communication systems is the repertoire of alarm calls uttered by the vervet monkeys of East Africa. Vervets lead a perilous existence, at constant risk from eagles, leopards and snakes, and they possess a distinctive warning call for each. When researchers record one of these calls and play it back to other vervets, the monkeys reliably scan the skies in response to the eagle call, look down at the ground at the snake call, and leap into bushes at the leopard call.

In an interesting link with human language, the basic mechanisms of the vervet's calls seem to be innate but are refined by learning. Baby vervets will give the eagle call in response to almost anything airborne, including falling leaves, but by the time they are adults the call has become focused on eagles, particularly the martial eagle, while nonpredatory birds like vultures are ignored.[35]

It is tempting to suppose that vervets have therefore developed a word for eagle, but that is not really the case. A vervet cannot combine two of its cries to state that, in its opinion, "Eagles are more dangerous than leopards!" Its calls can be used only as one-note alarms to warn that "An eagle is coming, take cover!" or "Leap—it's a leopard!"

Besides appearing to lack precise words for things, animals also lack the ability for syntax. Though capuchin monkeys seem to obey an ordering rule in their calls (for example, call A is made before calls B and C but never after them), the meaning of such ordered calls, if any, is not yet clear to researchers.[36] Strenuous efforts have been made to teach language to chimpanzees. The first attempts focused on training the chimps to make humanlike sounds. Then, when the unsuitability of their vocal apparatus was accepted, they were taught to communicate in sign language. Chimps can learn a number of signs—about 125, according to their trainers, more like 25, according to skeptics—but there is no consistent evidence that they use the order of the signs to confer meaning, as is the essence of human language. Typical utterances of Nim Chimpsky, a chimp trained by Herbert Terrace of Columbia University, were "Me banana you banana you me give," and "Give orange me give eat orange me eat orange give me eat orange give me you." "The chimp's abilities at anything one would want to call grammar were next to nil," concludes Steven Pinker of Harvard University.[37]

Still, evolution's design principle is continuity, so there must have been some neurological structure in the mammalian brain that was adapted to generate the combinatorial systems of vocabulary and grammar, just as the mammalian ear and voice box were adjusted to analyze human voices and generate human speech sounds. In an unusual alliance, the animal communication experts Marc Hauser and Tecumseh Fitch recently joined with the linguist Noam Chomsky to propose that the human capacity for syntax might have evolved out of an animal brain module designed for some other purpose, such as navigation.[38] Their argument is that the essential feature of language is recursion, the ability to embed one phrase inside another in an indefinitely long chain. Recursion may also be a feature of faculties like navigation that require an animal to remember how to get from A to D, with an excursion to B and C if the way is blocked. If the genes that specify the brain's navigation module were accidentally duplicated, the spare set would be free to evolve and perhaps acquire the function of encoding thought into language.

The Nature of Language

Many people think that thought would be impossible without language, and that the two are pretty much the same. Others equate language with speech. In the view of linguists, neither proposition is true. Animals may have quite rich thought processes—chimpanzees certainly know the position of all the individuals in their hierarchy and who must be recruited in a conspiracy— but are unable to put their thoughts into words. And speech is just one modality for language, which can also be written, or conveyed as signs, as in American Sign Language. Linguists regard this and other signing systems as proper languages with the same properties as spoken languages, including a fully developed syntax, or set of grammatical rules.

In the view of linguists, language is neither thought nor speech but rather a system for translating thought into a physical output, usually speech or writing. The brain behaves as if it were performing this translation process with a pair of combinatorial systems, one of which generates vocabulary, the other syntax.

The combinatorial system for vocabulary is a remarkable solution to a difficult problem. Many animal species communicate with a set of calls, each of which has a specific meaning. If the same principle were followed in human language, the calls after a certain number would start to merge into one other and become very hard to distinguish. But natural selection has somehow hit on a way of generating infinite variety, by basing the vocabulary system on a very small set of individually discrete sounds. The sounds can be joined in a limitless number of combinations, and any of these compound sounds can be arbitrarily associated with a meaning to form words. The system is called combinatorial because it is based on combining different elements to generate words.

The combinatorial system for syntax is tricky to describe because it seems to perform several different though related tasks. The original fix on it came from Noam Chomsky's insight that there must be a universal system in the human brain for allowing children to learn the grammatical rules of whatever language they hear spoken around them. Languages have many different rules of grammar, but all seem to be variations on the same model. Chomsky called this learning machinery Universal Grammar, but the phrase is also used to refer to the basic design underlying all grammars.

This proposal still attracts objections from researchers who believe the mind is a general purpose learning system, a blank slate with no preprogramming or genetically based circuitry dedicated to particular behaviors such as the faculty for language. It's certainly true that human behavior seems to be under conscious control to a far greater extent than is that of other animals. But equally it is clear that many behaviors in animals are genetically guided. In some animals, like the laboratory roundworm, biologists have already learned how to alter certain genes and induce a different behavior. It's reasonable to assume that there is a genetic basis for much human behavior, particularly such basic but highly complex faculties as learning a language or recognizing faces.

In the case of language, the combinatorial systems for vocabulary and syntax are so sophisticated that it seems unlikely an infant could quickly learn them from scratch. It would seem far more efficient for evolution to embed the general ability for learning language in the brain's neural circuitry. As Darwin observed, the ability to learn the spoken language seems instinctual, but the ability to write is not, which is why it must be learned so laboriously in school. In support of the view that the basic elements of language are innate, a human gene that seems fairly specific to language has recently been identified, as discussed below.

The fact that children around the world learn languages so easily, and at the same stage of development, points strongly to the unfolding of a genetic program as the children reach a certain age. Chomsky asserted that Universal Grammar was innate, and indeed the language-learning machinery seems to be one of the many developmental programs that are wired into the genes and scheduled to unfold at a given time.

But Chomsky and other theoretical linguists have been less interested in the question of what evolutionary stimulus might have prompted the evolution of language. No full length article about the evolution of language appeared in the Linguistic Society of America's journal *Language* until 2000. "Why linguists have tacitly accepted just such a self-denying ordinance should be a topic of some interest to sociologists of science," writes Derek Bickerton of the University of Hawaii, one of the few linguists to have explored the origin of language.[39]

Several leading linguists blame Chomsky for the neglect. His proposed system of Universal Grammar was such a complicated mechanism that his critics argued there was no way it could have evolved, since it would have been useless until the full structure was in place. This was a misguided criticism since evolution explains very well how enormously complex organs such as the eye or ear have evolved. Nonetheless, Chomsky, rather than debate the point, discouraged any discussion of evolution, several leading linguists now say. "Opponents of UG argue that there couldn't be such a thing as UG, because there is no evolutionary route to arrive at it," writes Ray Jackendoff. "Chomsky, in reply, has tended to deny the value of evolutionary argumentation."[40]

"To the extent that Chomsky has been willing to speculate on language origins at all, his remarks have only served to discourage interest in the topic among theoretical linguists. He has adamantly opposed, for example, the idea that the principles of UG arose by virtue of their utility in fostering the survival and reproductive possibilities of those individuals possessing them," writes Frederick Newmeyer, a linguist at the University of Washington, Seattle.[41] These two critics are not without standing; Newmeyer was president of the Linguistic Society of America in 2002, Jackendoff the following year.

Chomsky denies that he ever discouraged people from studying the evolution of language and says that his views have been misinterpreted. "I have never expressed the slightest objection to work on the evolution of language," he says. He outlined his views briefly in lectures 25 years ago but left the subject hanging, he said, because not enough was understood. He still believes that it is easy to make up all sorts of situations to explain the evolution of language but hard to determine which ones, if any, make sense.[42]

For outsiders looking in, it's hard to understand why linguists such as Newmeyer and Jackendoff would blame Chomsky for the entire profession's neglect of evolution, given that his colleagues, as independent academics, were presumably capable of thinking for themselves. However, Chomsky did have a significant impact on what others thought, says Steven Pinker, in part because of his intellectual stature and in part because of an aggressive style of debate that polarized the whole field.

"Why should one man's opinion count for so much?" Pinker asks. "The fact is that Chomsky has had, and continues to have, an outsize influence in linguistics. He has rabid devotees, who hang on his every footnote, and sworn enemies, who say black whenever he says white. This doesn't leave much space for linguists who accept some of his ideas (language as a mental, combinatorial, complex, partly innate system) but not others (the baroque and ever-changing technical details of his theory of grammar, his hostility to evolution or any other explanation of language in terms of its function)."[43]

Like other social scientists, linguists have not made a habit of looking to evolution for explanations, even though it is the bedrock theory of biology. Pinker was one of the first linguists to do so. With Paul Bloom, he wrote an influential article with a self-declared "incredibly boring" goal. Its purpose was to explain to linguists that, contrary to the views of Chomsky and the science historian Stephen Jay Gould, "human language, like other specialized biological systems, evolved by natural selection."[44]

Pidgins, Creoles and Sign Languages

One of the Chomskyans' problems with evolution of language, that language is too complex to have halfway steps, has been addressed by Derek Bickerton. Bickerton became interested in the subject through his study of a fascinating language phenomenon, the development of creoles from pidgins. Pidgins are languages of limited vocabulary and minimal grammar, usually invented by two populations who have no language in common. The children of pidgin-speakers do something very interesting: they spontaneously develop the pidgin into a fully fledged language with proper grammatical rules. These developed pidgin languages are called creoles.

It seemed to Bickerton, as he studied Hawaiian creoles, that their development offered an insight into the evolution of human language. The first language was pidgin-like, he suggested, consisting mostly of vocabulary, and syntax was grafted on later. Several possible remnants of this proto-language still survive. If children are not exposed to language in early childhood, when their Universal Grammar machine is switched on and primed to learn, they may never learn any language properly. This happens very rarely, in the case of feral children allegedly brought up by animals, or when pathological parents imprison their children in the house and refuse to speak to them. Genie, a 13-year-old California girl, was found in 1970 wandering the streets with her mother. The two had escaped from a house where Genie had been penned in a bedroom from the age of 18 months and denied conversation. After her rescue, intense efforts were made to teach her to talk, but she never acquired fully grammatical language. Her utterances were stuck at the level of sentences like "Want milk," or "Applesauce buy store."[45]

Even this primitive form of language could have been extremely useful to an early human society. Other possible echoes of the inferred proto-language can be heard in syntax-free utterances such as "Ouch!" or the more interesting "Shh!," which requires a listener.[46]

Recently linguists have developed a new window into the innate basis of syntax through a remarkable discovery—the detection of two new languages in the act of coming to birth. Both are sign languages, developed spontaneously by deaf communities whose members were not taught the standard sign languages of their country. One is Nicaraguan Sign Language, invented by children in a Nicaraguan school for the deaf. The other, Al-Sayyid Bedouin Sign Language, was developed by members of a large Bedouin clan who live in a village in the Negev desert of Israel.

The Nicaraguan case began when children were brought to a school for the deaf founded in 1977 by Hope Somosa, the wife of the Nicaraguan dictator.[47] Instructors noticed that the children had learned little from their

Spanish lessons but had developed a system of signs for talking to one another. Each generation of kids taught it to the next, and the language has rapidly evolved from a set of gestures into a sophisticated language with its own syntax.

The Al-Sayyid clan consists of some 3,500 people descended from a single founder who arrived 200 years ago from Egypt and married a local woman. Since the third generation, marriage within the clan has been encouraged, so there is a considerable level of inbreeding. Two of the couple's five sons were deaf, as are about 150 members of the community today. The clan's village is isolated in part by geography and even more by social barriers, since other Bedouin look down on them. Its deaf members did not go to school until recently and so were not exposed to either Israeli or Jordanian sign languages. They developed their own, and the language is also used by their hearing relatives to communicate with them.[48]

According to a famous story in Herodotus's *History*, an Egyptian king tried to ascertain the nature of the first language by isolating two children from birth and waiting to see in what tongue they first spoke. Study of the two new sign languages confirms that King Psammetichus's experiment was misconceived: it is not specific words that are innate, but rather the systems for generating syntax and vocabulary. Among both the Nicaraguan schoolkids and the Al-Sayyid clan, a spontaneous sense of syntax has developed, specifically the distinction as to whether a word is the subject or object of a verb.

In addition, the Al-Sayyid signers have developed the preference for a specific order of words in a sentence, that of subject-object-verb. The Nicaraguan children, in contrast, have developed the signed equivalents of case endings for words. Since these indicate whether a word is subject or object, word order is not so important and keeps changing with each cohort of children.

The apparently spontaneous emergence of word order and case endings in the two sign languages strongly suggests that the basic elements of syntax are innate and generated by genetically specified components of the human brain. The Al-Sayyid sign language has been developed through only three generations but some signs have already become symbolic. The sign for man is a twirl of the finger to indicate a curled mustache, even though the men of the village no longer wear them. Change is also brisk in Nicaragua. At first the children represented the number twenty by flicking the fingers of both hands in the air twice, says Ann Senghas, a linguist who has been studying their sign language for 15 years. But the sign was too cumbersome and has now been replaced with a form, signable with one hand, that looks nothing like a 20 but can be signed fast.[49]

Sign languages emphasize an often overlooked aspect of language, that gesture is an integral accompaniment of the spoken word. The human proto-language doubtless included gestures, and could even have started with gestures alone. Michael Corballis, a psychologist at the University of Auckland in New Zealand, argues that language "developed first as a primarily gestural system, involving movements of the body, and more especially the hands, arms and face."[50] Speech would have evolved only later, he believes, because considerable evolutionary change had to occur to develop the fine muscles of the tongue and other parts of the vocal apparatus.

Corballis's idea has several attractive features. It would explain why word and gesture are so well integrated, and why people even gesticulate when talking on the telephone despite the fact that their listeners cannot appreciate the performance. But critics of the idea note that gesture-based languages would be useless in the dark and that they require those conducting a conversation to be looking at each other all the time. Spoken language suffers from neither constraint.

Evolutionary Pressures for Language

Once language started, whether in the form of word or gesture or both, its further evolution would doubtless have been rapid because of the great advantages that each improvement in this powerful faculty would have conferred on its possessors. Even while still in its most rudimentary form, language would have made possible a whole new level of social interactions. Precise and unambiguous thoughts could at last be shared among members of a community, whether for making alliances, indicating intention, describing people and places, or transmitting knowledge. Moreover each small improvement in the overall system, whether in precision of hearing or articulation or syntax formation, would confer further benefit, and the genes underlying the change would sweep through the population.

But easy as it is to see how a simple form of language might have evolved into a complex one, that doesn't answer the question of what particular stimulus brought language into being in the first place.

Language now plays so many roles in human society that it's hard to arrange them in some hierarchy and say one role was the root and the others its branches. But evolutionary psychologists have come up with several interesting suggestions about the possible pressures for language to evolve. Robin Dunbar at the University of Liverpool in England has proposed a "social grooming" theory of language. He notes that monkeys and apes spend an inordinate amount of time grooming each other's fur. This activity, besides curbing parasites, serves to cement social relationships. But social grooming sets a limit on the size of a monkey group, because members will have no time to search for food if there are too many acquaintances whose fur must be rubbed the right way.

In practice, different monkey species spend varying amounts of time on grooming one another, up to a maximum of 20% of their waking day, and this is among species whose typical group size is about 50 members. The maximum time available for social grooming, Dunbar argues, has effectively capped the size of monkey social groups at 50 members. How then did the typical size of hunter-gatherer groups grow to 150 members, a number that would in principle require everyone to spend 43% of their waking hours on social grooming, or its human equivalent? Because of language, Dunbar suggests. Language is so much more efficient a way of establishing and confirming social bonds that the requisite amount of social grooming could be cut way back. In a wide range of human societies, it so happens, the amount of time people spend in social interaction, or conversation, is 20%. The driving force behind the evolution of language, in Dunbar's view, was the need to bond people in larger social groups.[51]

A quite different explanation has been advanced by the evolutionary psychologist Geoffrey Miller. He believes that sexual selection—Darwin's theory that the peacock's tail is the evolutionary product of peahens' choices—is what has driven the evolution of language. Just as the richness and symmetry of the peacock's tail signals its freedom from parasites, so eloquence and articulate speech signal the quality of an individual's mind, and will be highly favored by both men and women in their sexual partners. Language is a device that lets us learn about potential mates more thoroughly than any other method, Miller writes.[52]

The Dunbar and Miller hypotheses are both evocative and each may hold some measure of truth. But it's not clear if either really accounts for the richness and precision of language. Most adult speakers of English have a vocabulary of 60,000 words, though the top 4,000 words account for 98% of conversation. Does one really need 60,000 words, or even 4,000, for the purposes of social grooming, or even impressing one's inamorata? Miller's answer is that excess is the hallmark of sexual selection—once selection has started, the character under selection is taken to extremes, like the stag's enormous antlers.

But for linguists, the essence of language is meaning and communication, and it seems unsatisfactory to explain its evolution on any other grounds. Pinker argues one should take into account the new ecological niche that humans had moved into, which was in fact a knowledge-laden environment requiring a wealth of new information about plants and animals, about how to make tools and weapons, and about goings-on in one's own society. People's longer life span made it worthwhile to gather information and transmit it to one's children and grandchildren. "Language," Pinker says, "meshes neatly with the other features of the cognitive niche. The zoologically unusual features of *Homo sapiens* can be explained parsimoniously by the idea that humans have evolved an ability to encode information about the causal structure of the world and to share it among themselves. Our hypersociality comes about because information is a particularly good commodity of exchange that makes it worth people's while to hang out together."

Pinker concludes that know-how, sociality and language are three key features of the distinctively human lifestyle and that the three factors coevolved, each acting as a selective pressure for the others.[53]

It would be easier to pinpoint the most likely stimulus for the evolution of human language if one could identify when language emerged. Obviously the joint human-chimp ancestor did not speak, or chimps would too. And all human races can speak equally well, so that fully articulate, modern language must have evolved before modern humans left Africa. This means language would have emerged after 5 million years ago and before 50,000 years ago. Paleoanthropologists have made strenuous attempts to pin down the development of language through anatomy, by looking at the shape of the brain as implied by interior casts of old skulls, or features such as the hyoid, a U-shaped bone that supports the tongue muscles, and the hypoglossal canal, a passageway through bone for the nerve bundle that wires up the tongue muscles. But these studies have not yet brought a great deal of clarity to the problem.

Paleoanthropologists have tended to favor the idea that language started early, with *Homo erectus* or even the australopithecines, followed by slow and stately evolution. Archaeologists, on the other hand, tend to equate full-fledged modern language with art, which only becomes common in the archaeological record some 45,000 years ago. Their argument is that creation of art implies symbolic thinking in the mind of the artist, and therefore possession of language to share these abstract ideas.

Other archaeological facts favor a late start for language. To look at the rough stone tools of the Olduwan (made between 2.5 and 1.7 million years ago) they seem to be just chipped pebbles, made with no particular design in mind. But the tools of the Upper Paleolithic, which began 45,000 years ago, are precisely shaped and so well differentiated from each other that it seems plausible their makers had a different word for each, and therefore had language. "It is as though Upper Paleolithic flint workers were saying 'This is an end-scraper: I use it as an end-scraper, I call it an end-scraper and it must therefore *look* like an end-scraper,'" writes the archaeologist Paul Mellars. He argues that the makers' evident emphasis on the precise visual shape of their tools "is probably exactly what one would anticipate if Upper Paleolithic groups had a much more complex and highly structured *vocabulary* for the different artifact forms." Given their much cruder tool kit, the Neanderthals might also have had language, Mellars thinks, but with a much simpler vocabulary.[54]

If fully articulate modern language emerged only 50,000 years ago, just before modern humans broke out of Africa, then the proto-language suggested by Bickerton would have preceded it. When might that proto-language first have appeared? If *Homo ergaster* possessed proto-language then so too would all its descendants, including the archaic hominids who reached the Far East (*Homo erectus*) and Europe (the Neanderthals). But in that case the Neanderthals, to judge by their lack of modern behavior, appear never to have developed their proto-language into fully modern articulate speech. That might seem surprising, given the advantage any improvement in the language faculty would confer on its owner, and the rapidity with which language might therefore be expected to evolve. So perhaps the Neanderthals didn't speak at all.

Discovery of a Gene for Language

A remarkable new line of inquiry bearing on the origins of language has recently been opened up by the human genome project. This is the discovery of a gene that is intimately involved in many of the finer aspects of language. The gene, with the odd name of FOXP2, shows telltale signs of having changed significantly in humans but not in chimps, exactly as would be expected for a gene serving some new faculty that had emerged only in the human lineage. And, through the ability of genetics to reach back into the distant past, the emergence of the new gene can be dated, though at present only very roughly.

FOXP2 came to light through the discovery by Jane Hurst, an English geneticist, of an unusual London family whose existence she reported in 1990. The family consists of three generations. Of the 37 members old enough to be tested, 15 have a severe language deficit. Their speech is hard to understand, and they themselves have difficulty comprehending the speech of others. If asked to repeat a phrase like "pattaca pattaca pattaca," they will stumble over each word as if it were entirely new. They have difficulty with a standard test of the ability to form past tenses of verbs ("Every day I wash my clothes, yesterday I_____my clothes"; four-year-olds will say "washed" as soon as they get the idea). They have problems in writing as well as speaking. The affected members of the family have been given intensive speech training but mostly hold jobs where not much talking is required. "Their speech is difficult to understand, particularly over the telephone, or if the context is not known. In a group of family members it is hard for you to pick up the pieces of the conversation, which is difficult to follow because many of the words are not correctly pronounced," says Faraneh Vargha-Khadem of the Institute of Child Health in London.[55]

Some of the first linguists to study the affected family members believed their problem was specific to grammar but Vargha-Khadem has shown that it is considerably wider. Affected members have trouble in articulation, and the muscles of their lower face, particularly the upper lip, are relatively immobile.

It could be argued that their defect stemmed from some general malfunction in the brain, which was not specific to language. But the IQ scores of the affected members, though low, fell in a range (59 to 91) that overlapped with that of the unaffected members (84 to 119).[56] The core deficit, Vargha-Khadem concluded, is "one that affects the rapid and precise coordination of orofacial [mouth and face] movements, including those required for the sequential articulation of speech sounds."[57]

The affected members of the KE family, as it is known, have each inherited a single defective gene from their grandmother. They provide the results of an experiment that no one would even contemplate doing in humans, but which nature has performed nonetheless—what happens if you disable a critical speech gene? And the one disabled in the KE family seems to operate at such a sophisticated level that it looks as if it were one of the last genes to be put in place as the faculty of language was perfected.

In 1998 a team of geneticists at Oxford University in England set out to identify the defective gene by analyzing the genome of KE family members. Their method was to look for segments of DNA that the affected members shared and the unaffected lacked. The Oxford team soon narrowed the cause of the problem to a region on chromosome 7, the seventh of the 23 pairs of chromosomes in which the human genome is packaged. Within this region lay more than 70 genes, and it seemed that it would take several years to study each gene and see which one was responsible. But Hurst then turned up a new patient with the same rare set of symptoms. The patient, a boy, had a break in his chromosome 7 that disrupted one of the genes in the section the Oxford team was studying. It was an easy task to identify which of the new patient's genes had been broken. It was a gene known as forkhead box P2, or FOXP2 for short.[58]

The Oxford geneticists, Cecilia Lai, Simon Fisher and Anthony Monaco, then analyzed all 267,000 DNA units in the FOXP2 genes of the KE family members. In all the affected members, and in none of the normal members, just one of these letters was changed from a G to an A (the four different kinds of chemical units in DNA are known for short as A, T, G and C). The switch to an A at this site in the gene meant that in the protein molecule specified by the gene, a unit that should have been an arginine was changed to a histidine (proteins are made up of 20 different kinds of units, known as amino acids, of which arginine, and histidine are two).[59]

How could a single mutation in a gene cause such a wide range of effects? The FOX family of genes makes agents known as transcription factors, which operate at a high level of the cell's control system. The agents bind to DNA and in doing so control the activity, or transcription, of many other genes. FOXP2 is active during fetal development in specific parts of the brain, and the protein transcription factor it makes probably helps wire up these brain regions correctly for language. Brain scans of affected KE family members seemed normal at first glance but a more sophisticated type of scan has shown they have considerably fewer neurons than usual in Broca's area, one of the two brain regions known to be involved in language, and more neurons than usual in the other region, known as Wernicke's area.[60]

FOXP2 is an ancient gene, and even mice possess a version of it. If the human version of the gene is intimately involved in the language faculty, then the gene would be expected to have changed in some significant way in the human lineage. Svante Pääbo and colleagues at the Max Planck Institute for Evolutionary Anthropology in Leipzig, Germany, analyzed the sequence of the FOXP2 gene in mice, in the great apes, and in people from the major continents. Some genes change quite rapidly over evolutionary time but FOXP2, they found, is highly conserved. Chimpanzees and gorillas carry the identical version of the gene, which must be the same as that possessed by the joint ancestor of chimps and humans who lived 5 million years ago. That version makes a protein that differs in only one of its 715 units from the version carried by mice, which shared a common ancestor with humans 70 million years ago. This means that from 70 million to 5 million years ago, a span of 65 million years, the FOXP2 protein underwent only a single change.

But its evolution suddenly accelerated in the human lineage after the human and chimp lineages diverged. The human version of the FOXP2 protein differs in two units from that of chimps, suggesting it was subject to some strong selection pressure such as must have accompanied the evolution of language.

All humans have essentially the same version of FOXP2, the sign of a gene so important that it has swept through the population and become universal. By analyzing the variations in the FOXP2 genes possessed by people around the world, Pääbo was able to fix a date, though rather roughly, for the time that all humans acquired the latest upgrade of the FOXP2 gene. It was fairly recently in human evolution, and certainly sometime within the last 200,000 years, he concluded.[61]

Language is such a complex faculty that it must be mediated by a large number of genes and have developed in several stages. Given the observation that the KE family's deficit seems to be in the power of fluent, articulate speech, Pääbo thinks FOXP2 may have been one of the last genes recruited to the language function, perhaps the final step in the development of modern human speech. "Maybe it made the last perfection of language, made it totally modern," he says.[62]

Pääbo regards the dating of the gene as compatible with Klein's argument that modern language evolved very recently and was probably the spur to the human behavioral changes seen in the archaeological record 50,000 years ago.

Societies with two kinds of people, of greatly differing language abilities, may have existed during the evolution of language. As each new variant gene arose, conferring some improvement in language ability, the carriers of the gene would leave more descendants. When the last of these genes—perhaps FOXP2—swept through the ancestral human population, the modern faculty of language was attained.

4

EDEN

We can see, that in the rudest state of society, the individuals who were the most sagacious, who invented and used the best weapons or traps, and who were best able to defend themselves, would rear the greatest number of offspring. The tribes, which included the largest number of men thus endowed, would increase in number and supplant other tribes. . . . As a tribe increases and is victorious, it is often still further increased by the absorption of other tribes. The stature and strength of the men of a tribe are likewise of some importance for its success, and these depend in part on the nature and amount of the food which can be obtained. All that we know about savages, or may infer from their traditions and from old monuments, the history of which is quite forgotten by the present inhabitants, show that from the remotest times successful tribes have supplanted other tribes.

CHARLES DARWIN, *THE DESCENT OF MAN*

WITH THE DEVELOPMENT OF LANGUAGE, the process of human evolution in Africa reached a decisive stage. After 5 million years, the human lineage that split off from apes had emerged into a people quite similar in their form and faculties to those who live today.

This people, which can be called the ancestral human population, was probably the first to have possessed fully modern speech, and the last from which all people on earth are descended. Since it dispersed so quickly after its formation, it may have endured for only a few thousand years.

Not only did the ancestral population probably have a fleeting existence, it seems to have survived by the narrowest of margins. It lived sometime between 50,000 and 100,000 years ago, probably nearer to the 50,000-year mark. Between 60,000 and 40,000 years ago much of Africa was depopulated, and only in East Africa can archaeologists detect a human presence.[63] The reason may have been a long period of dry climate that shrank the forests and dried out the savannas. The ancestral population itself, geneticists estimate, shrank to as few as 5,000 people.

From this village-sized population, the world was peopled. And since people in societies around the world behave in much the same way, the principal elements of human nature must already have been present in the ancestral human population before its dispersal into Africa and the world beyond.

It would be of the greatest interest to know everything about the ancestral human population—its way of life, its social structure, the roles of men and women, its religion, the language that its members spoke. Not a trace of these first people has yet been found by archaeologists. Yet despite the total lack of direct evidence, a surprising amount can now be inferred about the ancestral human population.

Geneticists can estimate how large the population was and, by identifying its closest descendants, can point to where in Africa the ancestral population may have lived. They can even say something about the language the ancestral people spoke. And by analyzing the behaviors common to societies around the world, particularly the hunter-gatherers who seem closest to the ancestral people, anthropologists can describe how the ancestral population probably lived and what its people were like.

The Genealogies of Eve and Adam

Because everyone in the world is descended from the ancestral population, geneticists can infer some of its properties by analyzing the DNA of living people, and then working backward.

Two parts of the human genome are particularly useful for this purpose. One is the Y chromosome, the only chromosome possessed by men alone. The other is known as mitochondrial DNA. These are the only two parts of the genome that escape the shuffling of genetic material between generations. The shuffling, an evolutionary mechanism for generating diversity rapidly at each generation, means that almost all other parts of the human genome have a pedigree that is at present too complex to un-tangle.*

Unlike most pairs of chromosomes, the X and Y do not exchange seg-ments of DNA between generations (except at their very tips). This is to en-sure that the Y's most important gene, the one that makes a person male, never gets shuffled into the X chromosome. The Y chromosome is therefore passed down essentially unchanged from father to son, generation after gen-eration. Mitochondrial DNA escapes shuffling through a different process. Mitochondria, cellular components that generate chemical energy, are for-mer bacteria that were enslaved long ago by animal cells. They live in the main body of the cell, outside the nucleus that holds the chromosomes. When the sperm fuses with the egg, all the sperm's mitochondria are de-stroyed, leaving the fertilized egg equipped with only the mother's mito-chondria. Because of this arrangement, mitochondria are bequeathed unchanged from mother to child (and a man's mitochondria are not passed on to his children).[64]

In addition to their exempt status, the Y chromosome and mitochon-drial DNA each have a special and surprising property of uniqueness. All men in the world today carry the same Y chromosome, and both men and women carry the same mitochondria. All of today's Y chromosomes were in-

---

*A one-paragraph summary of human genetics: The human genome consists of 2.85 billion units of DNA in all, packaged in large, individual molecules known as chromosomes. A person inherits one set of 23 chromosomes from each parent, so that each cell of the body holds a total of 46 chromosomes. Before the eggs or sperm are generated, the number of chromosomes must be halved, since when egg and sperm unite it will double. But before the halving process, the germline cells make each chromosome inherited from the individual's father line up with its counterpart chromosome inherited from the mother. Each pair of chromosomes then swaps corresponding chunks of DNA with its counterpart, so that a new pair emerges, each of which is now a medley of maternal and paternal genes. Each member of the new pair is tugged to oppo-site sides of the cell, which then divides to generate eggs or sperm. A special feature of this process concerns the 23rd pair of chromosomes, known as the X and Y sex chromosomes. Be-cause the Y carries the male-determining gene, which must never be swapped into the X, the two chromosomes do not exchange genes, except at their very tips. Long ago, the Y chromo-some was the same length as the X, but it has shed genes because, through lack of the diversity generated by the swapping process, many of its genes fell into disuse. Sperm carry either an X or a Y chromosome, whereas eggs always carry an X. Fertilization creates individuals with an X-Y pair (men) or an X-X pair (women). A consequence of this process is that people carry separately in their cells the set of chromosomes inherited from their mother and father; it is only when they come to generate their own eggs and sperm that the maternal and paternal genes are as-sorted into new, recombined chromosomes.

herited from the same, single source, a Y chromosome carried by an individual male who belonged to, or lived slightly before, the ancestral human population. The same is true of mitochondrial DNA; everyone carries the same mitochondrial DNA because all are copies of the same original, the mitochondrial DNA belonging to a single woman.

The metaphor is hard to avoid—this is Adam's Y chromosome, and Eve's mitochondrial DNA. The ancestral human population, of course, included many Adams and Eves, indeed about 2,500 of each if the geneticists' calculations are to be believed. So how did it come about that just one man bequeathed his Y chromosome to the whole world and one woman her mitochondria?

It's a curious fact of genetics that one version of a gene, especially in small populations, can displace all the other existing versions of the same gene in just a few generations, through a purely random process called genetic drift. Consider how this might work among surnames, which are passed on from father to son just like Y chromosomes. Suppose a hundred families are living on an island, each with a different surname. In the first generation, many of those families will have only daughters or no children at all. So in just one generation, all those families' surnames (and accompanying Y chromosomes) will go extinct. Assuming no new male settlers arrive on the island, the same unavoidable winnowing will happen each generation until only one surname (and Y chromosome) is left.

This is what has happened with the human Y chromosome. Every Y chromosome that exists today is a copy of the same original, carried by a single individual in the ancestral human population. The Y chromosomes of all the other Adams have perished at some point along the way when their owners had no sons.

But despite all being copies of the same original, Y chromosomes are not identical. Over the generations, mutations—the switch of one of the four kinds of DNA units for another—have built up on the Y. The mutations are harmless but serve the invaluable purpose for geneticists of assigning the owners of Y chromosomes to different male lineages. The reason is that once a man has acquired a novel mutation in his Y chromosome, all his sons will carry that mutation, and no one else will. If one of the sons has a second mutation, all of his descendants will carry the two mutations. Each new muta-

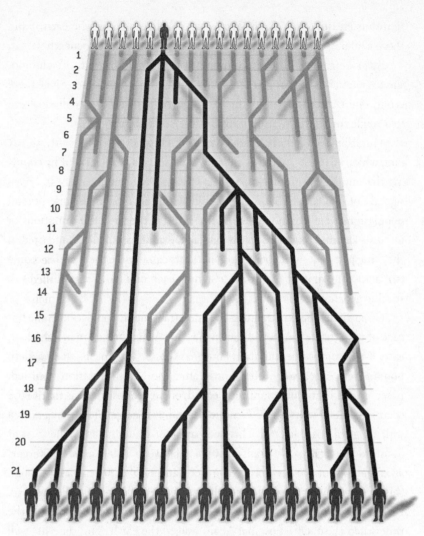

FIGURE 4.1. THE UNIVERSAL HUMAN Y CHROMOSOME.

Only men carry a Y chromosome. In each generation, some men have no children or only daughters, reducing the number of Y chromosomes in the population, until only one remains. This is why all men in the world carry a Y chromosome inherited from a single individual—the Adam of the Y chromosome—who lived in the ancestral human population. The same is true of mitochondrial DNA and the mitochondrial Eve.

tion thus creates a fork on the family tree—between those who carry it and those who don't—and stands at the head of all the lineages beneath it.

By looking at the most informative of the mutations on the Y chromosome, geneticists can assign every man to one lineage or another. Since there is only one Y chromosome, all these lineages or branches eventually coalesce to a single trunk, the Y chromosome of the original "Adam."

Mutations get incorporated into the Y chromosome at a fairly steady rate, which enables geneticists to put a date on each branch point by counting the number of mutations down a lineage. And the lineages can be assigned not only a date but a geographical location. This is because human populations were expanding across the globe at the time the mutations of interest occurred but then, to a remarkable extent, people lived and bred in the same place they were born. So geneticists can impose the Y chromosome tree across the map of the world, assigning each of its forks and lineages to specific geographical regions.

Of particular help in defining the ancestral human population is the lineage of men that left Africa. A few men inside Africa, and all men outside it, carry a Y chromosome mutation known as M168. This means that modern humans left Africa sometime shortly after the M168 mutation occurred. Based on the mutation-counting method, one recent estimate is that M168 occurred 44,000 years ago.[65] Genetic dates, however, generally come with a wide range of possible error. This one, say Peter Underhill and colleagues at Stanford University, could range anywhere between 39,000 and 89,000 years ago. The root of the Y chromosome tree dates to 59,000 years ago, though this too has a wide range of possibilities, from 40,000 to 140,000 years ago. Still, a date around 59,000 years ago seems a reasonable estimate for the time when the Y chromosomal Adam walked the earth. This date fits well with a date of 50,000 years ago for the ancestral human population, because genes tend to have slightly deeper ancestries than do populations.

Estimating the Ancestral Population Size

The Y chromosome is just a small part of the human genome. But it seems likely to represent human population history well enough, not least because its story is corroborated by mitochondrial DNA. Mitochondrial DNA can be

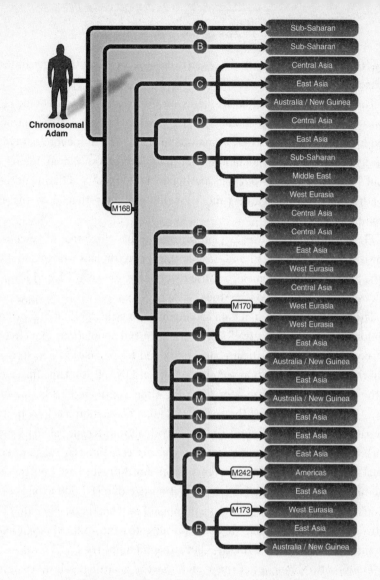

**FIGURE 4.2. THE Y CHROMOSOME FAMILY TREE
AND ITS GEOGRAPHICAL DISTRIBUTION.**

Although all men carry the same Y chromosome, mutations have gradually built up on it. The mutations allow men to be assigned to different lineages, depending on which set of mutations they carry. Because the mutations accumulated while the ancestral people were spreading through the world, different lineages of men are found in different regions of the world.

All male lineages outside sub-Saharan Africa carry the Y chromosome mutation known as M168. Men who carry the M173 mutation may have been the first modern humans to enter Europe 45,000 years ago, founding what archaeologists call the Aurignacian culture. Bearers of M170 are thought to have brought the Gravettian culture that succeeded the Aurignacian 28,000 years ago. The M242 mutation occurred just before the first humans crossed the Bering land bridge from Siberia to the Americas.

used to construct genealogies of women just as the Y chromosome generates lineages of men.

The women's lineages, like the men's, have all turned out to be branches from a single root, the mitochondrial DNA possessed by a single woman who lived in or before the ancestral human population. The mitochondrial Eve appears to have lived considerably earlier than the Y chromosomal Adam—about 150,000 years ago—but that may reflect the difficulty of dating mitochondrial DNA, which gathers mutations more rapidly than does the Y chromosome.

The mitochondrial genealogy of humankind has three main branches, known as L1, L2 and L3. L1 and L2 are confined to Africans who live south of the Sahara. The L3 branch gave rise to a lineage known as M, and it was the descendants of M who left Africa.

The Y and mitochondrial data can be made to yield another vital piece of information—the "effective" size of the ancestral population. The effective population is a statistical concept inferred by population geneticists from the amount of variation seen in samples of DNA. It is a large fraction of the real population size—for humans, often considered to be about half.[66] The effective size of the ancestral human population has long been estimated to have been around 10,000 individuals, but recent calculations, which mitigate a confounding factor in the earlier estimates, suggest the actual number may have been even smaller. An estimate based on the Y chromosome suggests an effective population size of just 1,000 men of reproductive age.[67] Assuming the same number of women, this implies an "effective population" of 2,000, which is equivalent to a census-size population of a mere 4,000 individuals, or say 5,000 in round numbers.

The first two branches of the Y chromosome genealogy, whose bearers are found only in Africa, have many sub-branches. This suggests the ancestral human population soon became quite spread out and diverse. There are other hints in the pattern of mutation that many Y chromosome lineages that once existed are now extinct. The ancestral population, in other words, may have suffered several calamities with widespread loss of male life.[68]

Because foragers lived in groups, generally of 150 people or so who may have liked to trade with neighboring groups, a population as small as 4,000 or 5,000 people is unlikely to have been distributed over the whole continent

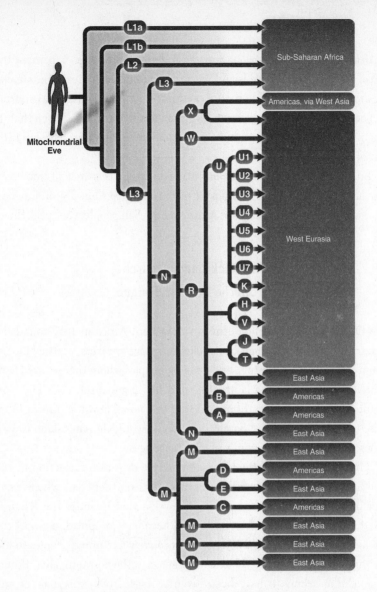

FIGURE 4.3. THE MITOCHONDRIAL DNA FAMILY TREE
AND ITS GEOGRAPHICAL DISTRIBUTION.

All men and women carry the same mitochondrial DNA, derived from a mitochondrial Eve. Like the Y chromosome, this single version of mitochondrial DNA has over time collected mutations which can be used to assign women to lineages (mitochondrial DNA is inherited only through the female line).

All women in sub-Saharan Africa belong to one of the first three branches of the mitochondrial DNA tree, known as L1, L2 and L3. All women outside Africa belong to M or N, the two daughter lineages of L3. Women in the western part of the Eurasian continent are all daughters of N; those of the eastern part descend from M or N. The daughter lineages A, B, C, D and X reached the Americas.

of Africa. It would probably have had a much smaller range—"perhaps the size of Swaziland or Rhode Island" according to one estimate.[69] The smaller the area, the more possible that at one time a single language was spoken. Archaeologists have not yet located this ancestral homeland. Given that its inhabitants would have been hunters and gatherers, they may have left little sign of their presence.

But geneticists have figured out where in Africa this ancient homeland may have been located. The clues lie not just in genes but also in tongues, specifically the click languages of Africa and the San people who speak them.

The Click Language Echo
from the Mother Tongue

The Dutch settlers who first arrived in southern Africa in 1652 found the country inhabited by two groups of click language speakers—cattle herders, whom they called Hottentots, and foraging peoples whom they referred to as either bushmen or San, a Hottentot word for "original settlers." The Hottentots called themselves Khoi-Khoi but are now known as Khwe. From these two words, Khoi and San, is derived the word Khoisan, which is used to describe the linguistic family of click languages.

The southern San of the Cape were largely driven to extinction by the Dutch settlers. Anthropologists have studied the northern San, who live in a large area from southern Angola to Botswana that includes the Kalahari desert. Until the 1970s, when settlement became widespread, many of the San lived as hunters and gatherers, one of the few remaining peoples to follow this ancient way of life. The main language of the northern San is !Kung, a name that seems to have been invented by German missionaries and means "they" in the Angola !Kung dialect.[70] These northern San are often referred to as the !Kung or the !Kung San. The "!" represents one of the many click sounds in their language.

To add to the confusion, anthropologists have recently started to refer to the !Kung San by their name for themselves, the Ju|'hoansi, which means "the Real People." The Real People's name for both Europeans and non-San Africans is !ohm, a category that includes predators and other inedible beasts.

The "|" in Ju|'hoansi designates one of the click sounds that are used as extra consonants in click languages. There are 5 kinds of click made by sucking air in, and a larger number made by expelling it. The "|" is an in-coming dental click, made by sucking the tongue in smartly from the upper front teeth, like the "tsk, tsk" sound used to indicate disapproval to children. The sound systems of Khoisan are said to be among the most complex in the world.[71]

About 30 different click languages are still spoken in southern Africa. They fall into three groups that, apart from their clicks, bear little evident relationship to each other. Speakers of another two click languages, known as Hadza and Sandawe, live far away in Tanzania. Hadza and Sandawe are both isolates, meaning they have no known relationship to each other or to the !Kung language of the San.

Despite the fact that many of the click languages apparently have nothing in common save their clicks, Joseph Greenberg, the great classifier of the world's languages, assigned them all to a single family, known as Khoisan. Linguists grumbled that it was illogical to define a group of unrelated languages as a family, but went along with the idea because no one knew what else to do with the click languages. Greenberg is at present reviled by most historical linguists, but his classification of the Khoisan languages seems a stroke of genius in light of a surprising new link that has now emerged among them.

The link is the finding from genetics that the Hadza speakers and the !Kung are two of the most ancient populations in the world. All peoples are of course the same age in the sense that everyone is descended from the ancestral human population. But some populations are viewed as older than others because they lie on longer branches of the human family tree. In a recent survey of African populations, Douglas Wallace of the University of California at Irvine found that three of the most ancient peoples in the world were the Biaka pygmies of the Central African Republic, the Mbuti pygmies of the Congo, and the !Kung San.

The !Kung possess several lineages of mitochondrial DNA, Wallace and his colleagues found, but their principal lineage forms the first branch of L1, the oldest of the three divisions of the human mitochondrial tree. This lineage, Wallace notes, "is positioned at the deepest root of the African phylogenetic tree, suggesting that the !Kung San became differentiated very early during human radiation."[72] In other words, the !Kung San split off from the ancestral population at an early date, and have remained a reasonably distinct population ever since.

Two Stanford researchers, Alec Knight and Joanna Mountain, recently compared the genetics of the !Kung with that of the Hadzabe, as the speakers of Hadza click language are known, a foraging people who live near Lake Eyasi in Tanzania. They discovered that the Hadzabe too are an extremely ancient people. However, the Hadzabe belong not to the L1 division of the mitochondrial DNA tree but to L2. Because L2 and L1 mark two of the first forks in the tree, the !Kung of L1 and the Hadzabe of L2 are two populations that separated almost at the dawn of human time. The split between the ancestors of the two groups "appears to be among the earliest of human population divergences," the Stanford researchers say. Based on measurements of their Y chromosomes, the two populations are more distant from each other than any other known pair of African populations.[73]

This genetic discovery provided a plausible reason why the two click languages, !Kung and Hadza, are also so different. Because the two peoples have been separate for so long, both their genetics have become very different and their languages have lost any resemblance to each other, save for the clicks.

The !Kung and the Hadzabe are both hunter-gatherers, and their nomadic lifestyle in the wilderness may explain how they have preserved their isolation from other groups for millennia. But that leaves the puzzle of why, when everything else in their language has changed, they have still retained the clicks.

Some linguists see this as a case of independent invention. They argue that there is nothing special about clicks, since a child can learn them, and that click sounds may have been lost and reinvented many times in language history just like the other features of language.

But on that theory clicks should be used in languages all over the world. In fact they are spoken only in Africa, with the exception of Damin, an extinct Australian language, of limited vocabulary, used for ceremonial purposes by the men of the Lardil tribe of Queensland. Clicks do not seem to have spread beyond their original speakers, with the exception that some click sounds have been borrowed by the San's Bantu-speaking neighbors. They are used for special purposes, such as in hlonipha, a respect language practiced by Nguni women to avoid the syllables of their in-laws' names; one way to avoid these taboo syllables is to substitute a click for one the consonants.

Though speakers of non-San languages may occasionally borrow or invent clicks, Anthony Traill, a click language expert at the University of Witwatersrand in South Africa, believes that "it is highly improbable that a fully fledged click system could arise from non-click precursors."[74] One reason is the difficulty of attaining fluency with multiple clicks. A single click is easy enough, but rattling off a whole series is another matter. Like double consonants, clicks are easy to stumble over. "Fluent articulation of clicks in running speech is by any measure difficult. It requires more articulatory work, like taking two stairs at a time," Traill says.

Another reason is that clicks seem easier to lose than to gain. In the ordinary process of language change, certain kinds of click can be replaced by non-click consonants, but Traill has never seen the reverse occur.

Given the laziness of the human tongue, why have clicks been retained by click speakers? "That is a major problem," Traill says. "All the expectations would be that they would have succumbed to the pressures of change that affect all languages. I do not know the answer."

If clicks are generally only lost, not invented, and if two of the oldest known populations in the world, the Hadzabe and the !Kung San, speak click languages, then it's possible that clicks were part of the first language spoken by the ancestral human population. "The divergence of those genetic lineages is among the oldest on earth," says Knight, the anthropologist on the Stanford team. "So one could certainly make the inference that clicks were present in the mother tongue."

Tracing the Boundaries of Eden

Since the San, on the basis of both genetics and language, seem to be among the earliest human populations, it's of considerable interest that they once occupied a much larger area of Africa than they do today. In the seventeenth century they inhabited all of southern Africa. Archaeologists believe that much earlier, in Paleolithic times, the San occupied the eastern half of Africa, with their domain stretching up through Ethiopia to the northern tip of the Red Sea.

In support of the archaeologists' view, geneticists have found that the DNA of Ethiopians living today retains evidence of the San's ancient presence in their country. Men of the Oromo and Amhara peoples have a small proportion of Y chromosomes that belong to the first branch of the Y chromosome family tree. This branch is rare elsewhere in Africa except among the San, 44% of whom carry it. The Oromo and Amhara must share an ancestral paternity with the San, and the first branch must have been "part of the proto-African Y-chromosome gene pool," writes Ornella Semino of the University of Pavia in Italy.[75] The mitochondrial DNA from women of the Oromo and Amhara peoples also indicates that Ethiopia, or at least east Africa, is the place from which the first modern humans left Africa.

Three major events in modern human evolution—the perfection of language, the formation of the ancestral human population and the exit from Africa—seem to have happened quite close to each other in time, around 50,000 years ago. The closer in time, the more likely that they happened in the same place and if so, Ethiopia seems at present the best candidate for being the birthplace of modern humans, the real world's counterpart of Eden's mythical garden.

Between the Universal People and the Real People

What were the people of the ancestral human population really like? Archaeologists describe them as "behaviorally modern humans," in contrast with the anatomically modern humans who first evolved nearly 200,000 years ago. But the term "behaviorally modern" refers to people whose traces in the archaeological record are not appreciably different from those left by contemporary hunter-gatherers. Foraging people of very different natures could leave much the same archaeological record.

It seems unlikely that the ancestral people closely resembled contemporary populations in behavior. The human skull and frame were then much heavier than those of people alive today, suggesting that the ancestral human population was physically aggressive, more accustomed to violence and warfare. Its members did not settle or build, perhaps because the social adaptations required for settled life were not yet part of their behavioral equipment. If fully modern language had evolved only recently, it is unlikely that all the other elements of contemporary social behavior emerged simultaneously with it. More probably they fell into place one by one as part of the continuing evolution of human behavior.

Yet the ancestral population, even if generally more inclined to aggression, presumably possessed all the major elements of human behavior that occur in its descendant populations around the world, since otherwise all of these behaviors would have had to evolve or be invented independently in each of thousands of societies.

There are two ways of developing a portrait of the ancestral human population; one is through the Universal People, the other through the Real People.

The Universal People is a concept of the anthropologist Donald Brown, who devised it as a counterpart to Chomsky's Universal Grammar. Though most anthropologists emphasize the particularity of the societies they study, Brown is interested in the many aspects of human behavior that are found in societies around the world. These universal human behaviors range from cooking, dance and divination to fear of snakes. Many, such as the facial expressions used to express emotion, seem likely to have a strong genetic basis. Others, like language, may result from the interaction of genetically shaped behaviors with universal features of the environment. Whatever the genesis of these universal behaviors, the fact that they are found in societies throughout the world suggests strongly that they would have been possessed by the ancestral human population before its dispersal.

These ubiquitously shared behaviors define the nature of what Brown calls the Universal People. Among the Universal People, families are the basic unit of social groups, and groups are defined by the territory they claim. Men dominate political life, with women and children expected to be submissive. Some groups are ordered on the basis of kinship, sex and age.

The core of a family is a mother and her children. Marriage, in the sense of a man's publicly recognized right of sexual access to a woman deemed eligible for childbearing, is institutionalized. Society is organized along kinship lines, with one's own kin being distinguished from more distant relatives and generally favored over those who are not kin. Sexual regulations constrain or eliminate mating between genetically close kin.

Reciprocity is important in the daily life of the Universal People, in the form of direct exchange of goods or labor. There are sanctions, ranging from ostracism to execution, for offenses such as rape, violence and murder.

The Universal People have supernatural beliefs and practice magic, designed for such purposes as sustaining life and winning the attention of the opposite sex. "They have theories of fortune and misfortune. They have ideas about how to explain disease and death. They see a connection between sickness and death. They try to heal the sick and have medicines for this purpose. The UP practice divination. And they try to control the weather," Brown writes.[76]

The Universal People have a sense of dress and fashion. They adorn their bodies, however little clothing they may wear, and maintain distinctive hairstyles. They have standards of sexual attractiveness. They dance and sing.

They always have a shelter of some kind. They are quintessential toolmakers, creating cutters, pounders, string to tie things together or make nets, and weapons.

The ancestral human population presumably possessed many, if not all, of the behaviors of the Universal People. It may also have had much in common with the San, who as members of the L1 branch of the mitochondrial tree may be the closest living approximation to the ancestral human population. Just how close is a matter of disagreement among social anthropologists. Some believe that little resemblance should be assumed between contemporary hunter-gatherers and those who lived thousands of years ago—people are always adapting genetically to their environment and there has been plenty of time for change. But foragers have presumably had much the same environment for the last 50,000 years. Chimpanzees seem to have changed very little in the last million years, so periods of evolutionary stability are not out of the question for human societies too. The lives of contemporary foragers are certainly not identical to those of early humans, but probably they overlap in many ways.

It was explicitly to help explore early human evolution that a group of Harvard anthropologists and others began, in the 1960s, a thorough study of the San, who still followed a foraging way of life. The choice of the San would have seemed even better had their ancient genealogy been known at the time. Unlike early hunter-gatherers, the San may have been confined to the less desirable regions of their former range, but even so they have little difficulty gathering enough food for their needs. They practice a foraging way of life that may have been typical of human existence ever since the days of the ancestral population.

Although the San's mitochondrial L1 lineage makes them only cousins to the people who left Africa for Asia (a sub-branch of mitochondrial L3), they bear some striking physical resemblances to Asian populations, suggesting that both lineages may have inherited these features from the ancestral human population. Many Khoisan speakers have yellowish skin, the epicanthic folds above the eyes that give some Asian eyes their characteristic shape, shovel-shaped incisors (front teeth hollowed out on the tongue side of the mouth, found commonly in Asians and Native Americans), and mongoloid spots—a bluish mark on the lower back of young infants. The !Kung San themselves apparently recognize this similarity since they assign Asians to the category of Real People like themselves, as distinct from !ohm, the category of non-San Africans and Europeans.[77]

As foragers, the San live off the land. Their principal food is mongongo nuts, but they are excellent botanists and recognize more than 200 species of plant, many of which they consider edible. According to Richard Borshay Lee's classic study of the !Kung San, some 60 to 70 percent of their food comes from plants they gather and 30 to 40 percent from meat gained by hunting. The !Kung are expert trackers. They can tell the species of animal that made a track and how many hours ago it passed; they can even identify individual people from their tracks. Their hunting bows are lightweight because they poison their arrow shafts with a lethal toxin. They obtain it from the pupae of any of three species of chrysomelid beetle that they dig up from beneath the bushes where the larvae have fed. The pupae stay in arrested development for several months, enabling hunters to carry them around and freshen up the toxicity of their arrows when needed. A well placed arrow will kill a 200-kilogram antelope in 6 to 24 hours.[78] In the laboratory, 25 trillionths of a gram of the arrow poison extracted from one of the beetle species, *Diamphidia nigro-ornata*, is enough to kill a mouse.[79]

Foraging life is neither as precarious nor as arduous as it might seem. Because of the diversity of resources the !Kung know how to tap, they cope easily with failures of supply by shifting from one source to another. Archaeological records suggest that their way of life in the Kalahari has persisted for thousands of years without a break.

It takes the !Kung 12 to 21 hours a week to gather all the food they need, according to Lee. Including other work activities like tool-making and maintenance, their total work week is 40 to 44 hours.[80]

The !Kung live in small groups that move camp whenever the surrounding food sources have been eaten out. A family's total possessions—tools, ostrich shell canteens, children's toys, musical instruments—pack into two bags. Nothing is stored, since everything they need is obtainable from the environment. Portability imbues !Kung life so thoroughly that it affects even the spacing of children. A woman can carry one child easily along with all her possessions, but two are a burden. !Kung women tend not to have a second child until the first can walk well. Children are not weaned until the age of four and before that age are carried almost everywhere, whether on foraging trips or when moving camp. Lee calculates that !Kung women walk about 1,500 miles a year, at least half of this distance carrying substantial burdens of food, water or possessions. A !Kung mother carries her child a total of 4,900 miles before it walks by itself.

Perhaps because a woman must invest so much care and labor in raising a child, she examines her newborn carefully for signs of defects. "If it is deformed, it is the mother's duty to smother it," writes the demographer Nancy Howell.[81] Infanticide is not the same as murder, in the !Kung's view, because life begins not with birth but when the baby is taken back to camp, given a name and accepted as a Real Person. "Before that time, infanticide is part of the mother's prerogatives and responsibilities, culturally prescribed for birth defects and for one of each set of twins born," Howell says. Women give birth outside the camp and men are excluded by taboo from the birth site; the reason for the taboo is doubtless that the father's absence makes easier the mother's decision as to whether to keep the newborn.

Land is owned collectively. Almost everything is shared, starting with the meat a hunter kills. Two character traits strongly discouraged by the !Kung are boasting and stinginess. Hunters are expected not just to distribute their kill but also to be extremely diffident about their success. This is central to the egalitarian ethos of !Kung society. Men in fact vary widely in their hunting skills, Lee found, but because they do not get to keep the extra meat or put on airs, not even the mightiest hunter can raise his social standing above others.

In the nineteenth century the !Kung used to live in groups with names like The Giraffes, The Big Talkers, The Scorpions, and even The Lice. One could only marry outside one's group. By the time of Lee's study, in the 1960s, the !Kung lived in more informal groups based around families and their near kin and in-laws. But the groups were small, around 30 or so people, and their size seems to have been limited by the nature of their sociality.

During the winter dry season, many groups would come together, to share goods, arrange marriages, hold feasts and do trance dancing. "But this intense social life also had its disadvantages," Lee writes. "The large size of the group required people to work harder to bring in food and fights were much more likely to break out in large camps than in small camps."

Because !Kung groups are strictly egalitarian, there is no authority to resolve conflicts and keep order. !Kung groups do have leaders, but they are informal, with no authority other than personal persuasion. The usual method of expressing disagreement is to vote with one's feet and leave camp along with one's family and followers. Lee noticed that large groups of !Kung stayed together for long periods only at the cattle camps of their Herero neighbors. The reason was in part "the legal umbrella provided by the Herero to maintain order among such a large number of feisty !Kung"—in other words, the Herero maintained a social order that the !Kung apparently had difficulty providing for themselves.

Attractive as an egalitarian foraging society may seem, it has certain drawbacks. Both private property and privacy are kept to a minimum. Without authority or a headman, individuals must resolve, by themselves or with the aid of kin, any disputes they cannot walk away from. And without specialized roles and some kind of hierarchy, a human society cannot grow beyond a certain level of size or complexity.

An early study of the !Kung, by Elizabeth Marshall Thomas, was titled *The Harmless People*.[82] The !Kung have many attractive qualities but harmless they are not. Fighting had been suppressed by the time of Lee's study, but rock art and historical accounts attest to its prevalence in the past.[83] The San fought regularly with their pastoralist Bantu neighbors, often raiding their stock and fending off counterattacks with their poisoned arrows. The Cape San in the Sneeuwburg mountains halted the expansion of the better armed and mounted Boers for 30 years until overwhelmed by the Boers' greater numbers.[84] As to internal violence, the !Kung's homicide rate, Lee found, is 29.3 per 100,000 person years, some three times that of even the United States.

Disagreements in !Kung groups escalate through the three recognized levels of talk, fighting and deadly fighting. The talk stage also has three sublevels. It starts as argument, moves up to verbal anger, and ends in *za*, a mode of pungent and personal sexual insult. These fighting words lead quickly to physical aggression. At that point, or shortly after, the poison arrows come out.

When hit with an arrow, the !Kung quickly cut around the wound and suck out the poisoned blood and lymph; chances of survival are 50-50. Puzzled at the high risks of this kind of conflict, Lee asked the naïve question of why men didn't use ordinary arrows instead. "To this question," he reports, "one informant offered an instructive response: 'We shoot poisoned arrows,' he said, 'because our hearts are hot and we really want to kill somebody with them.'"

The anthropologist gained another insight into !Kung methods of conflict resolution while conducting interviews about hunting success. Having asked four !Kung hunters how many giraffe and deer they had killed, Lee reports, "it suddenly occurred to me to pose the question: 'And how many men have you killed?'

"Without batting an eye, ≠Toma, the first man, held up three fingers; ticking off the names on his fingers, he responded, 'I have killed Debe from N≠amchoha, and N//u, and N!eisi from /Gam.'

"I duly recorded the names and turned to Bo, the next man. 'And how many have you killed?'

"Bo replied, 'I shot //Kushe in the back, but she lived.'

"Next was Bo's younger brother, Samk"xau: 'I shot old Kan//a in the foot, but he lived.'

"I turned to the fourth man, Old Kashe, a kindly grandfather in his late sixties, and asked: 'And how many men have you killed?'

"'I have never killed anyone,' he replied.

"Pressing him, I asked, 'Well then, how many men have you shot?'

"'I never shot anyone,' he wistfully replied. 'I always missed.'"[85]

Ancestral Portrait

It is tempting to suppose that our ancestors were just like us except where there is evidence to the contrary. This is a hazardous assumption. The ancestral human population is separated from people today by some 2,000 generations. In evolutionary time, that is not so long, yet is still time enough for very substantial evolutionary change to have taken place.

Consider that the anatomically modern humans of 100,000 years ago showed no signs of modern behavior. They had no apparent capacity for innovation and may have lacked the faculty of speech. Very significant evolutionary change seems to have occurred in the 50,000-year span that separates them from the behaviorally modern humans of the ancestral population. Yet that is the same span of time that separates the ancestral population from people today, allowing for an equally decisive evolutionary change. And the pace of human evolution may well have accelerated in the last 50,000 years, given the unparalleled changes in environment experienced by the ancestral people as they left their homeland, colonized strange lands and cold climates, and converted from foraging to settled life.

Indeed specific evidence has now emerged suggesting that the human brain has continued to evolve over the last 50,000 years. The evidence, as described in the next chapter, rests on the finding that two new versions of genes that determine the size of the human brain emerged only recently, one around 37,000 years ago and a second at 6,000 years ago. Given the brain's continued development, the people of 50,000 years ago, despite archaeologists' tag for them as "behaviorally modern," may have been less cognitively capable than people today.

The ancestral human population would have lived by hunting and gathering, and its way of life was perhaps not so different from that of foragers like the !Kung San. In its homeland in northeast Africa, the ancestral people were doubtless as skilled at exploiting the plants and animals of their local environment as the San are in theirs. They would have possessed a carefully thought-out suite of tools for hunting, food preparation, and carrying things. To judge by the journey of those who were to leave Africa, they probably knew how to build boats and how to fish.

But their technology would have been considerably less sophisticated than that of the !Kung. The !Kung's lightweight bows and poisoned arrows represent a high degree of mechanical and biological knowledge. There is no clear evidence that the bow was invented until some 20,000 years ago. It never reached Australia, suggesting it was not known to the ancestral human population.

Without projectile technology, male hunting success in early human societies would have been considerably less spectacular. Large animals would
have been hard to kill, so hunters perhaps concentrated on small game that
they could run down and spear. "Before effective hunting, males could have
focused more on honey and plant foods, so their daily hauls of food did not
have to be lower but must have been different," writes the evolutionary anthropologist Frank Marlowe. But women's foraging, for plant foods and tubers excavated with digging sticks, may have been much the same as in
contemporary foraging societies.[86]

In appearance, the ancestral human population would certainly have
had dark skin as protection against the African sun. They had stronger bones
and were thicker set than contemporary people. They would have cut and
decorated their hair. From the date assigned to the evolution of the human
body louse, which lives only in clothing, the ancestral people must have
worn clothes that were sewn to fit the contours of the body tightly enough
for the lice to feed.

It is tempting to suppose the ancestral people looked like the San, with
their lightish skin and slightly Asian cast of features, or perhaps like the aboriginal tribes of Australia, who have dark skin and wavy hair. But these two
groups have been evolving independently for 50,000 years and their appearance is unlikely to have remained unchanged. The ancestral people may
have been similar to both but would also have possessed their own distinctive appearance, which cannot at present be reconstructed.

The ancestral people spoke a fully articulate language, which may
well have included the click sounds still used by their Khoisan-speaking
descendants.

As hunters and gatherers, the ancestral people probably lived in small
egalitarian societies, without property or leaders or differences of rank.
These groups engaged in constant warfare, defending their own territory
or raiding that of neighbors. When they grew beyond a certain size, of
150 or so people, disputes became more frequent, and with no chiefs or
system of adjudication, a group would break up into smaller ones along lines
of kinship.

Yet these quarrelsome little societies would have contained in embryo
the principal institutions of the large modern societies of today. They had
some form of religion, a practice that seems as old as language and may have

coevolved with it. Religion may have served as an extra cohesive force, besides the bonds of kinship, to hold societies together for such purposes as punishing freeloaders and miscreants or uniting in war.

A sense of fairness and reciprocity governed exchange and social relationships. Much later, the idea of reciprocity would be extended to non-kin, allowing strangers to be treated as honorary relatives and creating the framework for societies that transcended the kin-bonded tribe.

Warfare may have been a dominant factor in the ancestral population's existence. A group could attain respite from conflict by finding new territory. Yet it could not have been easy to travel far from the ancestral homeland. Foragers are adapted to surviving in their local environment by their intimate knowledge of its plants and animals. Only one group of people, a little band maybe only a few hundred strong, succeeded in overcoming the daunting odds and leaving the homeland altogether. But by daring so much, they gained the whole world.

5

EXODUS

Let us suppose the members of a tribe, practising some form of mar-
riage, to spread over an unoccupied continent, they would soon split
up into distinct hordes, separated from each other by various barriers,
and still more effectually by the incessant wars between all barbarous
nations. The hordes would thus be exposed to slightly different condi-
tions and habits of life, and would sooner or later come to differ in
some small degree. As soon as this occurred, each isolated tribe would
form for itself a slightly different standard of beauty; and then un-
conscious selection would come into action through the more powerful
and leading men preferring certain women to others. Thus the differ-
ences between the tribes, at first very slight, would gradually and in-
evitably be more or less increased.

CHARLES DARWIN, THE DESCENT OF MAN

O
UT OF AFRICA there are two routes. One is to travel above the
northern tip of the Red Sea and through the Sinai desert into the
Levant, the lands of the eastern Mediterranean. The other is to cross
the Red Sea at its southernmost point, the Bab al-Mandab or Gate of Grief.

For the ancestral people in their East African homeland, the vast desert
to their north may have been a serious barrier. But the Gate of Grief was almost on their doorstep.

Today these straits are some 12 miles wide. But 50,000 years ago much
of the world's ocean water was locked up in the glaciers of the Pleistocene
ice age, and the straits would have been much narrower. At that date the water level of the Red Sea was some 230 feet lower than at present.[87] The
straits were never completely closed but at low sea-stands they would have
been dotted with islands, an inviting chain of stepping-stones to the southern Arabian peninsula.

"With the blast of thy nostrils the waters were gathered together, the
floods stood upright as an heap, and the depths were congealed in the heart

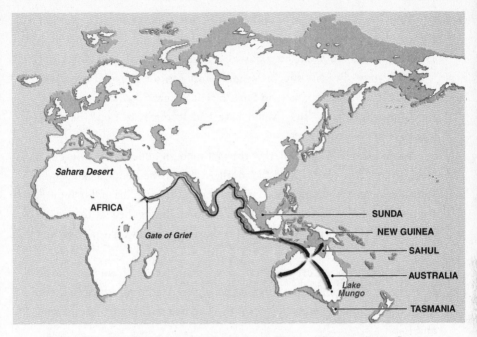

FIGURE 5.1. THE ROUTE FROM AFRICA TO THE FORMER CONTINENT OF SAHUL.

Humans left Africa at a time when ice sheets covered northern latitudes of Europe and Asia, and sea levels were some 200 feet lower than now. They probably crossed the Red Sea at its southern entrance and reached India. Generation by generation, people expanded along the coastlines of southern Asia until by 40,000 years ago they had reached the foundered continent of Sahul (now Australia, New Guinea and Tasmania).

This eastward route taken by the first modern humans to leave Africa may reflect a preference for staying within the tropical climates to which they were adapted, or the occupation of the mainland by *Homo erectus*, or both.

of the sea." Those were Moses' words of appreciation for the divine wind that parted the waters of the Red Sea, just in time for the Israelites to escape from Egypt and the pharaoh's pursuing chariot forces.[88] Too bad that the epic, if less miraculous, crossing of the first modern humans into the world beyond Africa cannot be reconstructed in equal detail. Still, some essential features of this ancient exodus are clear enough.

The first, based on genetic analysis, is that there seems to have been just a single emigration of modern humans from Africa. A second genetic inference is that the number of those who left was probably quite small. Indeed it could have been as few as some 150 people, raising the puzzle of why, if one group of people managed to escape from Africa, many more did not do so.

After some differences of opinion, geneticists now seem to agree that the trees drawn on the basis of the Y chromosome and of mitochondrial DNA both point to a single exodus from Africa. "Analysis of mtDNA and Y chromosome diversity support a single East African source of migration out of Africa," say two geneticists in a recent review.[89] If there were many migrations, they add, "they would have had to originate from the same source population in Africa."

It's reasonably likely, then, that the first modern humans left Africa in a single group, that they crossed the southern end of the Red Sea and slowly spread, generation by generation, around the coasts of Arabia and Iran until they reached India. Because of the lower sea levels during the Pleistocene ice age, the archaeological evidence of this coastal passage would now lie underwater.

But this version of events is not yet generally accepted. Another possibility, favored by some experts, is that people traveled from Africa to India by a northern route, across the top of the Red Sea and through the Levant and Iran.

A third possibility is that there were at least two migrations, one to the north and the Levant, the other to the south and India. This theory is favored by archaeologists who have reservations about the reliability of genetic inferences. But if the geneticists are right that there was only one migration, a choice must be made between the northern route, across the top of the Red Sea, and the southern route, across the sea's southern end; and the present weight of evidence, at least in the geneticists' eyes, favors a single exodus via the southern route.

In tracing the movements of the first modern humans across the globe, geneticists' maps show neat arrows stretching from eastern Africa to India, Australia or Japan, and the arrows unavoidably give the impression that the emigrants were purposefully traveling to these distant endpoints. But of course they were not—they had no maps and no idea of what lay at the end of their journey. In fact, it's doubtful they they were on a journey at all.

For foraging people, short journeys may be routine but long distance travel, carrying their infants and all necessities, is arduous. Rather than trek determinedly into the unknown, or expose their families to the hazards of exploration for its own sake, it's more likely that the first modern humans to leave Africa behaved as foragers usually do—they moved a short distance and stayed put. After a number of years, as new births swelled the group's size, it would have divided so as to prevent the usual discord that wells up in large foraging populations.

Following such splits, one group would stay put while the other moved on into unclaimed territory. Foragers need a lot of space to support themselves, so in a century—five generations—a hunter-gatherer society might spread over a considerable distance, especially if its members had learned the art of coastal fishing and preferred to stay near the water's edge. Those long distance migrations, in other words, were not made by a single group on a long trek, but were the slow expansion of human populations who took a generation to travel each leg of the journey.

Crossing from Africa to the Arabian peninsula via the Gate of Grief would have required boats. Though archaeologists have found no water craft from this period, people who lived at African sites of the Later Stone Age, which began shortly after 50,000 years ago, could certainly fish, so boat building techniques may well have been familiar to the ancestral human population. If the emigrants left Africa by boat, their descendants may thereafter have moved along coastlines until they reached India.

The coastlines could well have been safer than the interior. Southern Arabia is for the most part an inhospitable desert and would have presented a formidable obstacle to foragers. But from time to time on the geological time scale it enjoys rainy periods. Even within the Pleistocene ice age that gripped the world until 10,000 years ago, there were periods of relative warmth. One, known as oxygen isotope stage 3, peaked around 50,000 years ago. During this warm phase, as well as two earlier ones, southern Arabia was wetter and would have been habitable by hunter-gatherer populations.[90]

But these spells of favorable climate may also have drawn down Neanderthals from the north. The Neanderthals may have thwarted previous attempts by humans in Africa to cross into Arabia, just as they crushed the attempt by anatomically modern humans to penetrate the Levant. By 50,000 years ago, however, the Neanderthals would have faced a different adversary. The ancestral people, with their new gift of language, would have enjoyed better organization and superior weaponry. Though physically weaker than the Neanderthals, the new model of humans may at last have gained an edge over their fierce archaic relatives.

Still, having their families with them, they may well have preferred to keep out of the archaics' way. So instead of striking out across the interior, they may have expanded along the coastline of southern Arabia, using their boats both as transport and to fish from.

So why was there only one migration of modern humans out of Africa? Could it have been that there was only one way out—the Gate of Grief— and the first people to cross it stayed put on the other side and prevented others from invading their territory? Perhaps more likely is that the odds of survival were small, and only one group of people was fortunate enough to surmount all the daunting obstacles in their path.

Passage to India

How can the long ago journey of those first emigrants be traced? Because of the territorial behavior of the first modern humans, rigorously maintained as they invaded the world outside Africa, everyone essentially stayed in place in their new home, except for those at the head of the wave of advance. The world would thus fill up in a rather orderly way. For thousands of years there-after, people lived and died in the place where they were born. Populations did not mix, except at a local level under the patrilocal system of hunter-gatherer societies.

This conclusion emerges directly from the genealogies of the Y chromo-some and the mitochondrial DNA. Men of each branch of the Y chromosome tree are mostly found in a particular geographic region. The same is true for mitochondrial DNA lineages. Some branches of the two trees are confined to a single continent. Others spread over several land masses, but in a way that tracks an orderly movement of population. If the world's population were highly mixed, each of these branches would be found all over the place.

Even today, most people in the world still belong to Y chromosome and mitochondrial DNA lineages that accurately reflect their continent of ori-gin. Africans south of the Sahara belong to mitochondrial lineages L1, L2 and L3. All the rest of the world belongs to the two daughter lineages of L3 known as M and N.

Lineage M is of particular interest in tracking the exodus from Africa. Sil-vana Santachiara-Benerecetti of the University of Pavia in Italy has found that M is quite common in people of the southern Arabian peninsula, but is not seen in the Levant.[91] This is interesting evidence that the route out of Africa may have led through southern Arabia to India. It is not conclusive, however, because a vigorous Arab slave trade flourished between AD 650 and 1900 and brought many Africans to Arabia. The female slaves, who became integrated into Arab populations, could be the source of the M lineage.[92]

Since men and women had to spread through the world together, the story told by the Y chromosome's genealogy should match that of the mitochondrial DNA genealogy. The two stories do agree well in general outline, though not yet in every detail. The mutations that generate the Y chromosome's genealogy were discovered more recently and are still under study. Mitochondrial DNA mutates at a faster rate than the Y chromosome and the dates derived for the branch points in its genealogical tree are generally older than those for the Y chromosome's tree. It will take much more work to get dates of the two trees into correct alignment.

Besides helping to track the movement of the first modern human emigrants around the globe, the two trees also record that only a small sample of the African population emigrated to the rest of the world, a conclusion also implicit in the fact of a single migration. On the female side, lineages L1 and L2 of the mitochondrial DNA tree remained confined to Africa, at least until modern times; only the M and N daughters of L3 left for the world beyond. On the Y chromosome genealogy, sons A and B of the Y tree never left Africa. Only the group of sons carrying a mutation known as M168 are found outside Africa. Presumably men of the M168 lineage accompanied the M and N lineage women as they and their descendants migrated from Africa to India.

The emigration of modern humans from Africa was not only a watershed in history but also a significant demographic event. The few who left Africa carried only a small subset of the genetic diversity present in the ancestral human population. Genetic diversity refers to the number of alternative versions of each gene—known as alleles to geneticists—that exist in a population; each individual can carry up to two of these alleles, one inherited from each parent. For instance, a region of DNA associated with the insulin gene exists in 22 different versions in African populations, but only three of these occur outside Africa.[93] The small size of the departing population would have increased the chances of its following a different evolutionary path from the host population in Africa because it created the conditions for the important kind of evolutionary change known as genetic drift.

Natural selection is the better known agent of evolutionary change but drift is also powerful, and the smaller the population, the more quickly drift acts. The mechanism of drift is the purely random way in which about half of a parent's genes get passed on to a child and half are discarded.* Depending on the luck of the draw, some versions of a gene become more common in a population as one generation succeeds another, while others grow rarer. Eventually one version of a gene may become universal while all the alternative versions are lost. This is what has happened in the cases of the Y chromosome and mitochondrial DNA. In terms of evolution's overall process, drift is the counterpart to mutation. Mutation constantly injects novelty into the genome, and in each generation drift sweeps novelty away. Natural selection draws on this flux, using it to keep each species adapted to the changing environment.

Since drift is random, the versions of a gene that it makes universal may be good or bad. For the most part, though, they are neutral, in geneticists' parlance, because they make no difference to an organism's survival. The smaller a population, the fewer generations it takes for a particular version of a gene to become universal. So drift would have been enhanced among the small group that left Africa and in its far-flung descendants as they spread out across the world. The human population as a whole probably existed for many millennia as small, largely separate groups, because distance and territoriality would have deterred any substantial mixing of peoples.

Those who left Africa carried only a slice of the full genetic diversity of the human population, and the size of the slice allows an estimate to be made of the emigrants' numbers. Sarah Tishkoff, a geneticist at the Univer-

---

*Everyone carries about half of their father's genes and half of their mother's; so what happened to the half you didn't inherit? It gets discarded, along with all the genes it contains. Each gene comes in a variety of different versions, known as alleles. By sheer chance, proportionately more of some alleles may get into the next generation, just by the luck of the draw, while fewer fall into the discard pile. The frequency of a given allele in the population may thus change considerably from one generation to the next, from 5%, say, to 33%, to 13%, to 55%, and so forth in a random fashion. But this random walk cannot go on forever. Sooner or later the frequency of the allele in the population will hit one of two numbers, 0% or 100%. At 0%, the allele is lost forever from the population. At 100% it becomes universal, i.e., the only version of that gene in the population since all other alleles are lost. When an allele becomes universal, geneticists say it is "fixed." The time it takes for any allele to become fixed depends on the number of generations and the size of the population, being faster when the population is smaller. When an allele becomes fixed, the population is then set on a different path through evolutionary space than if the other alleles of that gene had remained available to it.

sity of Maryland, has calculated that the number of modern humans who left Africa could have been as few as 160.[94] Another estimate, made by geneticists working with mitochondrial DNA, is that the source population in Africa from whom all humans outside Africa are descended numbered at most 550 women of childbearing age, and probably considerably fewer.[95]

Despite the appearance of precision, these numbers have wide ranges of error and are very approximate. The basic inference that can be drawn from them is that the ancestral group in Africa from which the first emigrants derived was very small, probably just a single band of hunter-gatherers. Such a band would number about 150 people if modern hunter-gatherer groups are typical of ancient ones. The group that left Africa would presumably have been this one band or a part of it.

Peopling the Lost Continents of Sunda and Sahul

By whichever route the first humans left Africa, India seems to have been their first major stopping point, because it is there that are found the first diversifications, outside Africa, of the mitochondrial and Y chromosome trees.

In terms of the mitochondrial DNA tree, the M and N lineages that came out of Africa are still frequent in today's Indian population. The M lineage is very common, and its mutations are older than those of M lineages found farther east, supporting the idea that the Indian subcontinent was settled soon after the African exodus. On the Y chromosome side, several offshoots of the early male lineages are restricted to the Indian subcontinent, a finding consistent with the scenario that the first settlers arrived by a southern route; those offshoots would be expected to occur in the Levant as well as India if the emigrants had taken the northern route out of Africa.[96]

In India there was a historic parting of the ways. Some people continued the coast-hugging, population budding process along the southern shores of Asia, eventually reaching the Australian land mass, China and Japan. Others pushed inland in a northwesterly direction, through the lands that are now Iran and Turkey, and began the long contest with the Neanderthals for the possession of Europe. Both paths tested the power of the new modern people to innovate, survive in hostile surroundings, and overcome daunting obstacles. Consider first the migration to Australia, then the push into Europe.

The group expanding along the coast pushed eastward around India and Indochina, eventually reaching the two lost continents of Sunda and Sahul. With sea level much lower 50,000 years ago than it is now, the Malay peninsula and the islands of Sumatra, Java and Borneo formed a single land mass known as Sunda or Sundaland, which was a southern extension of the Asian land mass. Australia was then connected to New Guinea in the north and to Tasmania in the south, the three islands forming the lost continent known as Sahul, directly south of Sunda.

Apart from fording river mouths, the people expanding along the coastline would not have had to cross open sea until they reached the channel between Sunda and Sahul. This would have been a formidable barrier, some 60 miles wide. Forest fires in Sahul, or the flights of birds, may have indicated the presence of land to watchers from Sunda. In any event, modern humans reached Sahul, an achievement that puts beyond doubt their possession of the seafaring skills required to cross the Gate of Grief as a way out of Africa.

The arrival of the first modern humans in Australia is an important event, not just because they had accomplished an epic migration from their distant homeland in East Africa, but also because it offers one of the first opportunities to link genetic history with archaeological history. On the basis of burials, archaeologists believe Australia was settled shortly after 50,000 years ago. This period is beyond the reach of the radiocarbon method of dating, so an alternative method must be used, known as thermoluminescence. The method is not always reliable but in this case is supported by independent evidence: by 46,000 years ago, all large Australian mammals, birds and reptiles weighing more than 220 pounds had suddenly fallen extinct.[97] The reason was almost certainly the activity of a vigorous new predator, human hunters. The large animals of the Americas were to undergo a similar extinction shortly after the first hunters reached the New World.

It is perhaps surprising that Australia should hold the earliest archaeological sites outside Africa in which the presence of modern humans has so far been established. The likely reason is the comparative ease of migrating along the coastline instead of venturing inland. The sea route provided a reliable source of food and an easy means of travel, save for crossing the Sunda-Sahul passage. Since sea level was then much lower and many former coastal sites are now submerged, that could explain why no intermediate stages of the journey have yet come to light.

Many geneticists believe that the first modern humans came out of Africa considerably earlier than 50,000 years ago. Dating based on the rate of mutation seen in the tree of human mitochondrial DNA suggests that modern humans first left Africa 65,000 years ago, according to a recent calculation.[98] But genetic dates, though always interesting, depend on many assumptions that may not be realistic, and the dates derived by archaeologists are considerably more reliable. Archaeologists can at present see no sign of modern human presence outside Africa before 46,000 years ago, the date of the Lake Mungo site in southeastern Australia, and sites of similar age in the Levant. They have little patience with the geneticists' proposals that sites of earlier occupation along ancient coastlines now lie beneath sea level, since rather than wait to be engulfed by slowly rising sea levels, people would surely have built new settlements farther inland.

It's possible, of course, that the modern humans leaving Africa really were confined to the water's edge by the archaic humans who had settled Eurasia many thousands of years beforehand. The Neanderthals may have been present at times in the Arabian peninsula and *Homo erectus* occupied East Asia. Though the archaic humans may at first have been able to prevent modern humans from penetrating Eurasia, forcing them to skirt the periphery, the archaics themselves never reached Sahul. That could perhaps account for the odd fact that the oldest modern human remains come from the place at the remotest part of the journey, Australia.

Another reason for Australia being the first recorded landfall, however, is climate. The ancestral population was not adapted to northern climates. As discussed further in chapter 6, people may have needed to evolve special adaptations to colonize the colder regions of Eurasia. The first emigrants may have been confined by climate to the coasts of East Asia and warmer regions like Sunda and Sahul.

Be this as it may, the archaeologists are probably correct in their position that modern humans should not be assumed to have left Africa any earlier than 50,000 years ago, a date that is consistent both with the behavioral changes evident from archaeological sites within Africa, and with the date, 46,000 years ago, at which modern humans were clearly present in Australia.

But if archaeologists are right on the date of exit, geneticists may have the better case on the number of migrations: just one.

Because Sahul lies so far off the beaten track, away from the subsequent movements and mixings of the human population, Australia's aboriginal tribes may hold in their genes a fascinating portrait of the first emigrants from Africa. But differences in the robusticity of early skeletons, and the arrival of people with a semi-domesticated dog (the dingo), suggest that several extra waves of immigrants reached Australia after the first one. Because of political constraints on taking samples from aboriginal peoples, it's at present impossible to sort out these various waves of early immigrants as fully as geneticists would like.

Before the arrival of Europeans in Australia, the original inhabitants were divided into some 600 tribes, each composed of some 500 to 1,000 people and possessing its own dialect and territory. These tribes seem to have married within themselves, with little gene flow between them, and because of their antiquity each built up a distinctive genetic profile with special variants not seen elsewhere in the world. An analysis of mitochondrial DNA from the Walbiri tribe of the Northern Territory showed they possessed several lineages not found among any other tribe, indicating a considerable degree of genetic differentiation between them.[99]

In contrast to the diversity of mitochondrial DNA types, there are far fewer Y chromosome variations, with half of all male aborigines carrying one with the same distinctive genetic signature. This may result from what geneticists call a founder effect—the reduced genetic diversity of populations founded by a small number of individuals.[100] Other factors may also have been at work. Polygamy, when some men have many wives and others none, is a powerful reducer of diversity among Y chromosomes. So too is frequent warfare, the burden of which is borne by men.

Australian aboriginal tribes seem to have lived in a state of constant warfare, with defended territories and neutral zones marked for trading. Their tool kit, designed for easy transport over long distances, included weapons like heavy war clubs, a special hooked boomerang, and spear-throwers.[101] The tribes were skillful at surviving in a harsh environment but never developed agriculture. Their special genetics reflects both their antiquity and the effects of genetic drift, promoted by the fragmentation of their population into small warring societies.

Genetic analysis has yielded similar insights into the lifestyle of another of Sahul's early occupants, the people of New Guinea. Australia and New Guinea were joined until about 8,000 years ago, so the peoples of both places may be descendants of the same migration. Mark Stoneking and colleagues analyzed both the Y chromosome and mitochondrial DNA from people in many New Guinean tribes and found a striking lack of diversity of Y chromosome lineages, especially in the highland tribes of the Dani, Yali, Una and Ketengban.[102] As with the Australian aborigines, reduced diversity could mean either a high degree of polygamy, with just a few men fathering most of a community's children, or a high rate of death in battle.

Both factors seem to have been at work in New Guinean society. All Papuan speaking populations in New Guinea practice patrilocality, with the men staying in their native clan and the women moving to their husband's clan. Most, if not all, New Guinea tribes practiced polygamy, at least until the missionaries arrived. Among the Dani, for example, 29% of the men had more than one wife, the range being from two to nine, while 38% of the men were not married.

Warfare was common in most Papuan societies until the second half of the twentieth century, Stoneking and his colleagues note, and casualty rates were high—about 29% of Dani men were killed in warfare, according to the anthropologist Karl Heider. This death rate is very similar to the male battle casualties among both chimpanzees and the Yanomamo of South America and presumably is driven by the same motive, the reproductive advantage gained by the successful warrior for himself and his male kin.

Warfare among hunter-gatherers is deceptively mild compared with the explosive carnage of modern battlefields. Battle may be opened but called off, like a ball game, if rain stops play, or someone is seriously injured. Heider, like many anthropologists, believed at first that warfare among the Dani was not a terribly serious affair. After his first field trip to New Guinea in 1961 he wrote a book entitled *Grand Valley Dani: Peaceful Warriors.* But after revisiting the Dani for many years, and reconstructing careful genealogies and causes of death, he realized how many men in fact died in battle. If you fight every week, even low casualty rates start to mount.

Like the !Kung San, the Dani fight to kill. They have not discovered how to daub their arrows with a poison like that of the chrysomelid beetles, but they use excrement instead, hoping to cause infection. Like many other human groups and the chimpanzees of Gombe and Kasakela, the Dani know that killing a few of the enemy leaves the remainder thirsting for revenge, so a more effective solution is extermination.

"About 30 percent of all independent highland social groups become extinct in each century because they are defeated," the archaeologist Steven LeBlanc writes of New Guinea tribal warfare. "These groups are either massacred or killed, or the survivors of a particularly deadly encounter flee and take refuge with trading partners or distant relatives. This last place on Earth to have remained unaffected by modern society was not the most peaceful but one of the most warlike ever encountered."[103]

The physical appearance of Australian aborigines is termed australoid, meaning that they have dark skin, wavy or curly hair, slender body build and large teeth. New Guineans are australoid but with minor differences, such as tightly curled hair. That the people of Sahul should look somewhat like sub-Saharan Africans is probably no accident. Because of their relative isolation, australoid peoples may be closer to the first emigrants than are most other living people. But they cannot exactly represent the first modern humans who left Africa because their population includes later immigrants, such as Polynesians, and they have themselves changed a lot through genetic drift.

The Enigma of the Andaman Islanders

Australian aborigines are not the only trace population left from the original migration. All along the route back to Africa, in remote islands or out of the way places where later invaders could be resisted, there are unusual peoples whose genetics suggest an ancestry from the original emigrants. All are tribal, mostly forest-living groups who have managed to resist intermarriage or integration. They include some of the tribal peoples of India, such as the australoid Chenchus and Koyas of Andhra Pradesh, as well as the Negritos, forest dwellers found in the Andaman Islands, Malaysia and the Philippines. Many of these peoples have dark skin, as if retained from their African origins.

The Andaman Islanders are one of the most intriguing of these relict populations. The Andaman Islands lie in the Bay of Bengal, some 120 miles from the coast of Burma, but with the lower sea levels of 50,000 years ago the distance may have been as little as 40 miles. Since the first emigrants from Africa were capable mariners, as proved by their reaching Sahul, the Andaman Islands would also have lain within their reach.

The islands were long avoided by contemporary sailors, their occupants having a fearsome reputation for extreme hostility and cannibalism. According to a British survey in 1858, the islands were inhabited by some 13 different tribes, each with its own language and territory, and some in a state of perpetual warfare with each other. Many of the northern tribes, known as the Greater Andamanese, were decimated by contact with western diseases, and within 50 years of British occupation almost all had perished. Only three of the peoples, all from the southern islands, now survive. They are the Onge, the Jarawa and the Sentinelese.

The origin of the Andamanese has long been a puzzle. Their features—short stature, dark skin, peppercorn hair and protruding buttocks, a feature known as steatopygia—are characteristic of African pygmies. "They look like they belong in Africa, yet here they are sitting in this island chain out in the middle of the Indian Ocean," says Peter Underhill, an expert on Y chromosome lineages. "People have been scratching their heads for 200 years asking who are these people and where do they come from."[104]

To address the question, two teams of researchers recently analyzed the islanders' DNA. Erika Hagelberg of the University of Oslo worked with blood samples from the Onge and Jarawa; she also extracted mitochondrial DNA from hair samples that had been collected from the Greater Andamanese by the ethnographer Alfred Radcliffe-Brown from 1906 to 1908.[105] A second team, led by Alan Cooper of the University of Oxford, obtained mitochondrial DNA from a collection of Andamanese skulls in the Natural History Museum in London; the ancient DNA was extracted from the pulp of teeth.[106]

Both teams found that the Andamanese belonged to the M2 mitochondrial lineage, and infer that they were part of the early migration of humans from Africa into southern Asia. The Y chromosomes of the Onge and Jarawa confirm the view that the Andamanese are an ancient, Asian people.

Their physical similarities with the African pygmies seem therefore to be what biologists call a convergent feature, meaning one acquired by independent evolution. Presumably when people start to live in forests, there are advantages in developing particular characteristics like short stature and steatopygia. The Biaka pygmies of the Central African Republic and the Mbuti pygmies of the Congo belong to different mitochondrial DNA lineages and presumably evolved pygmy stature independently of each other and the Andamanese.

With their dark skin and other African features, the Andamanese and other australoid peoples may represent what the early inhabitants of East Asia and Europe looked like before being displaced many thousands of years later by people from northern latitudes.

Another clue to the great age of the Andaman Islanders comes from language. Like the ancient !Kung and Hadza click languages, the Andamanese languages are isolates, meaning they are unlike each other and unlike any known language. The linguist Edward Sapir is said to have told his students that the world's languages are divided into two classes, Andamanese and all the rest.[107] This distinctiveness is another sign of great antiquity.

Joseph Greenberg, in his classification of the world's languages, placed Andamanese in a superfamily he called Indo-Pacific. The other members of Indo-Pacific are Tasmanian and the ancient Papuan languages of New Guinea. Like several of Greenberg's classifications, Indo-Pacific is not widely accepted by other linguists. But the grouping can now be seen to have put together languages that have another striking feature in common—all are spoken by people in remote regions who may be descendants of the first migration of modern humans from Africa to the foundered continent of Sahul.

The Penetration of East Asia and Indonesia

Australia was not the only destination for the first settlers of Asia. While some people crossed the straits from Sunda to Sahul, others presumably continued eastward around the southern borders of Sunda. They would have followed the coastline northward, up the eastern coast of China until they reached Japan and the Kamchatka peninsula, leaving a trail of settlements in their wake.

These groups, finding the coastlines in either direction inhabited, would eventually have started to push inland. They would have used rivers as highways into the interiors of India, Indochina, China and Central Asia, according to a reconstruction by the medical geneticist Stephen Oppenheimer. "Geography and climate decided the newly arrived occupants of Asia where to go next," he writes. "The rules would have been simple: stay near water, and near reliable rainfall; when moving, avoid deserts and high mountains and follow the game and the rivers."[108]

The penetration of the Eurasian land mass would have brought modern humans into direct conflict with the archaic humans who had long possessed it, certainly with the Neanderthals in the west and perhaps with *Homo erectus* in the east. Possibly this invasion was delayed for many generations until the innovative moderns had developed the necessary weapons and tactics to defeat the archaics or perhaps, less dramatically, until they had evolved the genetic adaptations for living in cold climates. The interaction between these different human species is of the greatest interest, but so far there is little data to go on, except the stark fact that one survived and all others perished.

In the east, for lack of archaeological studies, it is not yet known how widespread were the populations of *Homo erectus*, or whether in fact their disappearance had anything to do with the advance of the moderns. But the two human species did overlap in various ways, according to two quite unexpected pieces of recent evidence. The first comes from that intimate observer of human evolution, the human body louse.

David Reed, a louse specialist at the Florida Museum of Natural History, has found that people around the world carry two distinct groups of body lice that look alike but have genetically different histories. He made the discovery by constructing genealogies of the lice's mitochondrial DNA, just like other geneticists have done for people. But whereas all human mitochondrial DNA falls on the branches of a single tree, the louse DNA falls into two separate clusters. One of the clusters matches the human mitochondrial DNA tree both in date and geographical distribution, just as would be expected if the lice had divided into separate populations like their human hosts after the dispersal from Africa. The second cluster of louse DNA coalesces with the first but only in the distant past, some 1.8 million years ago, as if it had been living for most of the time on a different host.

Lice are highly specialized organisms and human lice cannot live for more than a few hours away from the warmth and sustenance of the human body. So this second cluster of lice must have been living on humans; it's just that they were of a different species, Dr. Reed believes. He suggests that they traveled out Africa with the ancestors of *Homo erectus* and much later switched across to the modern humans who came into physical contact with the *Homo erectus* populations in Asia some 50,000 years ago.[109]

A second and even more astonishing overlap between modern humans and

Homo erectus was recently reported from the Indonesian island of Flores, which lies between Indonesia and Australia. From the floor of a riverside cave, archaeologists recovered a series of fossil human remains of which the oldest is 95,000 years and the youngest 13,000 years. The remains belong to some seven individuals and include one complete skull. These people stood about three and half feet tall but were not human pygmies. Rather, they were a downsized version of *Homo erectus*, according to their discoverers and other experts.[110]

Island geography imposes special evolutionary constraints on arriving species, often propelling small species to giant size and downsizing large ones. The island of Flores was home to a species of giant rat and to lizards that evolved into the carnivorous Komodo dragons, 10 feet in length, as well as an even larger lizard, now extinct. This lost world was roamed by packs of pygmy elephants. And its human occupants too, it seems, were also downsized.

The little Floresians present many paradoxes with which paleoanthropologists are still grappling. They made sophisticated stone tools similar to those crafted by modern humans and unlike any previously associated with *Homo erectus*. Yet their brains, miniaturized along with their bodies, were about the same size as those of chimpanzees and the australopithecines, neither of which could fashion stone tools. Skeptics suggest that if the Floresians made the tools found with them, they must be modern humans, perhaps of some pathological form. But other experts say the surviving skull is clearly of *erectus* descent and shows no sign of pathology.[111]

On present evidence it seems that the little Floresians were descendants of *Homo erectus* who managed to endure some 35,000 years into the modern era, long after the rest of their species had perished. They owed their survival to living unobtrusively in a forest on a remote island. The only way that *erectus* could survive in the modern era, it seems, was by becoming essentially invisible to the new arrivals.

The Long Struggle against the Neanderthals

Unlike the still fragmentary evidence about the fate of *Homo erectus*, much more is known about the interactions of modern humans with the Neanderthals, the archaic humans who occupied Europe and the Near East. The Neanderthals, who evolved west of the Urals some 127,000 years ago, were a strikingly distinct variation on the human theme. Their bodies were stocky, with barrel chests and muscles like weightlifters'. They had large heads, with bony brow ridges on the front of their skulls, and strange buns or ridges on the back.

These special features may have been either a biological adaptation to cold, or the result of genetic drift, the random change in gene frequencies between generations. Genetic drift is especially powerful in reshaping small populations, as the early Neanderthals may have been.

Neanderthal remains include many broken and healed bones, suggesting their lifestyle was physically taxing—whether because of hunting game or each other is hard to say. Some skeletons bear injuries so severe that their owners seem likely to have depended on others to survive, suggesting that Neanderthals looked after their sick. They also, on occasion, practiced cannibalism, to judge by the cut and burned bones found at several sites. In both their pleasant and less pleasant behaviors, in other words, they were quite human.

Their brain size covered the same range, and in some cases exceeded, that of modern humans.[112] But their behavior was quite different. They used the same unvarying tool kit as anatomically modern humans, the forebears of the behaviorally modern people. They buried their dead in shallow graves, but there is no indisputable evidence that the burials were accompanied by ritual. At the Shanidar cave, in northwestern Iraq, a skeleton exhumed with large amounts of pollen pleasantly suggested floral tributes from fellow Neanderthals. But until any similar burial is found, the simpler explanation is that the pollen was imported by the rodents whose burrows honeycombed the grave fill.[113] There is some evidence that the Neanderthals were less socially cohesive.[114] Although they seem to have displaced anatomically modern humans from the Near East 100,000 years ago, they were unprepared for the highly innovative behavior of the humans who arrived on their doorstep 45,000 years ago.

"It is not difficult to understand why the Neanderthals failed to survive after behaviorally modern humans appeared," writes the paleoanthropologist Richard Klein. "The archaeological record shows that in virtually every detectable aspect—artifacts, site modification, ability to adapt to extreme environments, subsistence and so forth—the Neanderthals were behaviorally inferior to their modern successors, and to judge from their distinctive morphology, this behavioral inferiority may have been rooted in their biological makeup."[115] It is impossible to tell from their skeletal remains whether or not Neanderthals could speak, but the crux of their behavioral inferiority may have lain in their possessing only a crude, syntax-free protolanguage, or perhaps no language at all.

Some anthropologists have argued that the first modern humans may have interbred with Neanderthals. Given the hostility of human hunter-gatherer societies toward each other, and the extreme fear that the Neanderthals seem likely to have evoked in modern humans, it is hard to imagine that the two species enjoyed hanging out with each other, let alone that they would welcome an exchange of marriage partners. The human mitochondrial DNA and Y chromosome trees each coalesce to a single ancestor in Africa, with no sign of a Neanderthal contribution in either lineage.

The genetic separateness of Neanderthals was emphasized in 1997 in a dramatic feat of research by Matthias Krings and Svante Pääbo, then of the University of Munich in Germany. They managed to extract mitochondrial DNA from the original specimen of Neanderthal, some 40,000 years old, which was found in the Neander valley near Düsseldorf in 1856.[116] The DNA of the chromosomes in the cell's nucleus degrades quickly after death but the little ring of mitochondrial DNA, with about 1,000 copies in each cell, has a better chance of surviving for long periods. The extraction of the DNA was a technical tour de force, which many others had attempted but failed to do, in part because the method for amplifying DNA is prone to increase not just the target DNA but, even more so, the contaminating samples of human DNA that abound in every laboratory and handled object.

The Munich team managed to decipher only a small segment of mitochondrial DNA but enough to show that it differed significantly in its sequence of DNA units from that of modern humans. Mitochondrial DNA has now been extracted from a total of four Neanderthal fossils, situated in Germany, Russia and Croatia. All have DNA similar to each other and different from that of modern humans. Pääbo and colleagues have shown that Neanderthal mitochondrial DNA also differs from that of early modern humans, which weighs against the likelihood that Neanderthals made some mitochondrial genetic contribution to the modern human gene pool that has since been lost.[117]

But mitochondrial DNA represents only a small fraction of the genome, so the possibility that some Neanderthal genes may have been incorporated elsewhere in the genome cannot at present be ruled out.[118] Though a large scale intermingling of the two populations seems highly unlikely, modern humans may on occasion have enslaved and interbred with Neanderthal women. If so, Neanderthals, being adapted to the cold, would doubtless have had several useful genes to offer to modern humans and traces of these may yet be found even though the mitochondrial lineages have gone extinct.

Krings and Pääbo estimate that the mitochondrial ancestress of humans and Neanderthals lived 465,000 years ago, give or take a couple of hundred thousand years either way. Genes usually split sometime before populations split, so this means Neanderthals split away from the hominid line sometime after 465,000 years ago. Their presumed predecessors, known as *Homo heidelbergensis*, are known in Europe from around 500,000 years ago, but it is not until 127,000 years ago that distinctive Neanderthal fossils appear.

The Neanderthals' home territory stretched from Spain in the west to points east of the Caspian sea. In the Near East it included the lands that are now Turkey, Iraq and Iran. Perhaps modern humans first entered Neanderthal territory directly from Africa. But if, as suggested above, there was only a single emigration, the one that reached India, then modern humans would have arrived in Neanderthal territory by a route that led from northern India through Iran and Turkey. These invaders reached the Near East about 45,000 years ago and, according to the archaeological evidence, moved steadily across Europe.[119]

As the moderns advanced, the Neanderthals became restricted to peripheral refuges such as the Italian and Iberian peninsulas. With one puzzling exception, the Châtelperronian culture of 40,000 years ago, the Neanderthals stuck to their unchanging Mousterian tool kit, never learning from the innovative technology of their successors.[120]

There is no way to know for certain the nature of the interaction between the two human species. It is unlikely to have been pleasant. Hunter-gatherer societies cannot support standing armies, so it is probably wrong to think of the modern human entry into Europe as a military campaign. It was more a slow infiltration. Given that Pleistocene Europe had no highway system, the new arrivals may well have traveled by boat, along the northern coast of the Mediterranean and up the rivers of central Europe.[121] In winter, the frozen rivers would have made natural footpaths through the wilderness.

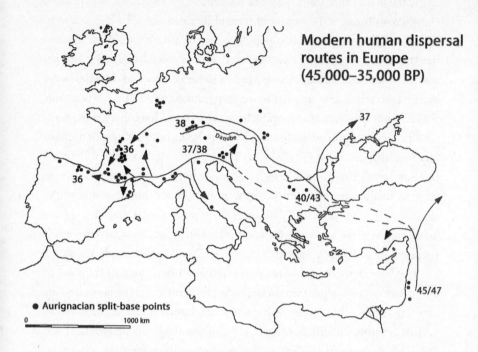

FIGURE 5.2. THE ARRIVAL OF MODERN HUMANS IN EUROPE.

Some of the African emigrants who reached India expanded to the northwest, through Iran and Turkey, eventually reaching Europe. Their slow-motion occupation of Europe took some 15,000 years, because of resistance from the indigenous Neanderthal population. Dots show sites occupied by Aurignacians, the name given by archaeologists to the culture of the first modern humans. Dates, in thousands of years before the present, are from radiocarbon measurements, and may be 3,000 years or so younger than calendar dates.

The modern humans probably moved as they always did, expanding into new territory as communities split, not exploring for the sake of adventure. Each new community would have skirmished with the local Neanderthals, who perhaps risked being killed by their fellows if they retreated into the territory owned by neighboring clans and had to hold on to their own territory or face extinction. Year by year, the moderns' territory expanded and the Neanderthals' shrank. From the extraordinary length of the process—a border war that took 15,000 years to move across Europe—it is evident that they did not yield easily. But by 30,000 years ago the Neanderthals had disappeared from their final refuges in the Iberian peninsula.[122]

The Moderns' Conquest

With the extinction of the Neanderthals, the archaics had been driven from the Eurasian land mass. Only the little Floresians survived, hidden in the forests of their remote island home. Modern humans, in the 20,000 years since their ancestors crossed the Gate of Grief, had occupied much of the world. Their populations, though still sparse, stretched across Eurasia, Sunda, Sahul and Africa.

But this was no imperium on which the sun never set, just a patchwork of tribes with no long range communication and no central authority. Archaeologists have found no towns or villages from this period; people still lived in a state of nature, wholly dependent on hunting and gathering for their existence.

For much of the period during which the exodus from Africa unfolded, from 50,000 to 30,000 years ago, people everywhere may have looked pretty much the same. Everyone outside Africa was descended from the 150 emigrants, who in turn were drawn from the host population in Africa.

The first modern humans were an African species that had suddenly expanded its range. For many millennia people would presumably all have had dark skin, just as do the relict populations of Australia, New Guinea and the Andaman Islands. It seems likely that the first modern humans who reached Europe 45,000 years ago would also have retained black skin and other African features. The Neanderthals, on the other hand, may have lived in northern climates long enough for the melanocortin receptor gene, which controls skin color, to have reverted back to its default state of producing pale skin. Though there exists no direct evidence as to skin color, and the point is only a curiosity, the Neanderthals may have had light skin and their conquerors black. Early Europeans, including the great artists of the Chauvet cave in France, may have retained the dark skin and other badges of their African origin for many thousands of years.

But despite the initial unity of the far-flung human family, regional differences inevitably arose. For archaeologists, the most striking are artistic. There is nothing to match the great painted caves of Europe, even though rock art of the same era is also known from Australia. "We must wonder," writes the archaeologist Ofer Bar-Yosef in discussing the art of this period, "why western Europe and, in particular, the Franco-Cantabrian region is so different from the rest of the Upper Paleolithic World. It is not the lack of limestone caves or suitable rock surfaces that prevented other social groups or their shamans from leaving behind similar paintings and engravings. Possibly this local flourish had to do with the vagaries and pressures faced by foragers in two major refugia regions at the ends of the inhabited world—western Europe and Australia—where there are claims for rock art of the same general age."[123]

There was a significant difference, or the seeds of a difference, between the European and Australian antipodes of the modern human advance from Africa. The Australian and New Guinean branch soon settled into a time warp of perpetual stagnation. They were still living with Paleolithic technology when their European cousins came visiting 45,000 years later. They never broke free from the triple bonds of patrilocal society, nomadic mobility and tribal aggression. For some reason the modern people who reached Europe and the Far East were able to escape this trap and to enter on a phase of steady and continued innovation.

Why these different modes of development occurred is one of the more puzzling questions of prehistory. Historians and social scientists, from the nature of their disciplines, tend to offer purely cultural or environmental explanations for all human differences. From a biologist's point of view, however, it seems likely that genetic influences would also have been at work, not least because it is hard to prevent an organism from responding genetically to a persistent environmental challenge. When people inhabit polar regions, they adapt genetically to the cold by developing the physique of Eskimos. When people go to live in tropical forests, they may develop pygmy stature, a change that has occurred independently at least three times since the diaspora from the ancestral homeland. Dispersed in small populations from Africa to Australia, from East Asia to Europe, the people of the Upper Paleolithic would have been subject to different evolutionary pressures and to the random effects of genetic drift.

Striking proof of the human tendency to develop local genetic variations has recently emerged from Iceland, whose population has been thoroughly studied by geneticists looking for the roots of disease. Iceland has been settled for just 1,000 years, by settlers from Norway, Britain and Ireland. Yet distinctive genetic variations have already arisen in each of eleven localities in

Iceland, according to a test developed by DeCode Genetics, a gene-finding company based in Reykjavik. The reason is that Icelanders, like people throughout the world, have tended to live, marry and die in the same place, and distinctive genetic variations have had time to develop in each locality, even in just 1,000 years. By scanning a person's genome, DeCode's researchers can specify where in Iceland that individual's parents and grandparents came from. The test is based on analyzing the sequence of DNA units at just 40 sites along the genome.[124]

If a detectable degree of local genetic differentiation has developed in Iceland in a mere 1,000 years, much greater differences are likely to have arisen among populations in the rest of the world, much of which has been settled for 40 times longer and where there have been many social and geographic impediments to the free flow of genes.

Genetic differentiation would certainly have started to act on the human populations of the Upper Paleolithic era. Bruce Lahn, a geneticist at the University of Chicago, has made a striking discovery about the evolution of two genes involved in the construction of the human brain. Each gene has several alternative versions, or alleles, but in each case one specific allele has become much more widespread than the others in certain populations. For an allele to rise to high frequency very quickly is a signature of natural selection hard at work. So presumably each allele conferred some very strong selective advantage.

One of the alleles is an alternative version of a gene known as microcephalin. The allele appeared around 37,000 years ago (though anytime between 60,000 and 14,000 years is possible) and is now carried by some 70% of many populations of Europe and East Asia. The allele is much less common in sub-Saharan Africa, where it is typically carried by from zero to about 25% of the population.

Just some 6,000 years ago a new allele of another brain gene, known as ASPM, appeared in the Middle East or Europe and rapidly rose to prominence, being carried by about 50% of people in these populations. The allele is less common in East Asia and occurs hardly at all in sub-Saharan Africans.[125]

What made the two alleles spread so quickly? It seems likely that each conferred some cognitive advantage, perhaps a slight one yet enough for natural selection to work on.

In Lahn's view, many genes are likely to be involved in constructing the human brain. He has found alleles of two of these genes, both of which happen to be quite common in Europeans and East Asians, but there almost certainly exist alleles of other genes that may be more common in other populations. Each population may therefore have used a different set of alleles to accomplish the same purpose, a well known biological process known as convergent evolution.

Resistance to malaria, for instance, is mediated by protective alleles in a number of genes, but Africans are protected by one set of alleles and Mediterranean peoples by a different, though often overlapping, set. The reason is that new alleles arise by mutation, a random process, and each population must make use of whatever alleles it has available. An advantageous allele may spread over time to neighboring populations, but will be more common in the place where it first arose. Lahn believes he is seeing the same phenomenon with alleles that have increased cognitive powers, and has just chanced on two alleles that happen to be common in European and Middle Eastern populations. "It is likely that different populations would have a different make-up of these genes, so it may all come out in the wash," he says.

Perhaps because of the sensitivity of suggesting that one population might have become genetically more acute than another, several critics asserted that the alleles could have become more common for some reason having nothing to do with the brain, such as conferring resistance to disease.[126] But there is at present no evidence that the microcephalin or ASPM genes do anything other than determine brain size. Some genes do play more than one role, but no other functions have yet been detected for microcephalin or ASPM. Their role in the brain, however, is well established. They first came to light because they are disabled in people with microcephaly, causing the brain to be much smaller than usual, particularly in the cerebral hemispheres that are the site of the brain's higher cognitive functions.

This strange condition seemed a throwback to the time 2.5 million years ago when the human brain was a third of its present size. In 2004 Lahn established that microcephalin and ASPM, along with several other brain genes, had undergone far more rapid evolution along the line of descent from monkeys to humans than had the counterpart genes in rodents.[127] The finding suggested that the brain has grown larger because a succession of new and more powerful versions of genes like microcephalin and ASPM were favored by natural selection. The most recent alleles of microcephalin and ASPM are just a continuation of this process, in Lahn's view.

A firm conclusion from Lahn's finding is that human evolution continued after the dispersal of the ancestral population 50,000 years ago, and took different forms in different populations. Much of this evolution may have been convergent, as each population adapted with different alleles to the same challenges. But convergent evolution does not necessarily proceed in lockstep in each separate population. So it could be that the spread of the microcephalin allele some 37,000 years ago expanded the cognitive powers of Caucasian populations and underlay such striking cultural advances as the Aurignacian people's adeptness at painting caves, while other populations developed such capabilities later.[128]

When the ancestral human population dispersed across the world 50,000 years ago, evolution set in motion a grand experiment: each population, in its fiercely guarded territory, would develop in its own way. This development would be both cultural, leading to a vast family of different languages, religions and lifestyles, and also genetic, as the members of each society responded to different climates, ecologies and social arrangements of their own making. Isolated on their separate continents, the far flung branches of the human family were to follow different trajectories as each adapted to the strange world that lay beyond the boundaries of their ancestral homeland.

6

STASIS

*Nomadic habits, whether over wide plains, or through the dense
forests of the tropics, or along the shores of the sea, have in every case
been highly detrimental. Whilst observing the barbarous inhabitants
of Tierra del Fuego, it struck me that the possession of some property,
a fixed abode, and the union of many families under a chief, were the
indispensable requisites for civilisation. Such habits almost necessi-
tate the cultivation of the ground; and the first steps in cultivation
would probably result, as I have elsewhere shewn, from some such accident as the seeds of a fruit-tree falling on a heap of refuse, and producing an unusually fine variety. The problem, however, of the first
advance of savages towards civilisation is at present much too difficult to be solved.*

CHARLES DARWIN, THE DESCENT OF MAN

B Y 30,000 YEARS AGO, with the Neanderthals ousted from Europe,
and *Homo erectus* confined to a relict population in the forests of Flores, modern humans had replaced the archaic people who had occupied the world outside Africa for more than a million years. But in a sense,
nothing had changed. The more modern humans, like the archaics whom
they evicted, were foragers who lived off nature's bounty. They built nothing
and left almost nothing behind, save for their stone tools. The newer humans crafted tools of far greater sophistication, including many of bone, and
made works of art such as ivory figurines and the decorated caves of France
and Spain. But they were still nomads, barred by their mobile way of life
from all the material and intellectual possibilities of civilization.

These hunter-gatherers had one more great transition to make before
entering the history of civilization. They had first to abandon their nomadic
way of life and settle down in fixed communities. Given the great advantages of settled life, presumably settlement was not a previously available
option. But why not? What made it so impossible for early people to put a

roof over their heads and enjoy the comforts of a fixed abode? What had to happen to make the transition to settlement possible? Why did ancestral humans need to spend 35,000 years wandering in the wilderness before conceiving the benefits of settled life and civilization?

Just as a physical change leading to pygmy form has often evolved in people who live in forests, so a behavioral change may have been necessary for people to abandon the nomadic life they had always known. Settlement may seem a natural choice to us, but it requires a set of wrenching adjustments for hunter-gatherers. They must learn to live with strangers. They must abandon the freedom to move away from danger or from people they don't get along with. They must yield their firmly egalitarian way of life for a hateful social order of superior and inferior, rife with rules and priests and officials.

Whether or not some genetic change was required to make it happen, the development of settled societies was a transition of profound importance. The interval from 50,000 years ago to 15,000 years ago, when the first settlements appear, is a formative period in human history, even though the precise road to settlement remains obscure.

This pre-settlement period is the subject of the present chapter, and the developments from settlement to agriculture are covered in the next. This division of the human past rests on the assumption that settlement was a decisive step in human evolutionary history and of considerably greater significance than one of its consequences, the expansion of agriculture some 10,000 years ago. Such a demarcation, however, cuts across the usual division recognized by archaeologists. They call the period from about 45,000 to 10,000 years ago the Upper Paleolithic age, based on its characteristic suites of stone tools. These gave way 10,000 years ago to tools of the Neolithic age, which is also equated with the beginning of agriculture.

The shift from the Upper Paleolithic to the Neolithic occurred at the same time as a major climatic transition, the ending of the great Pleistocene ice age that began 1.8 million years ago and the beginning of the Holocene, the warm period that has lasted to the present day. As with the Upper Paleolithic age, the end of the Pleistocene epoch is also set at around 10,000 years ago.

With Africa and Australia already inhabited, the principal developments in human history during the period from 50,000 to 15,000 years ago were those taking place in the Eurasian land mass. Though Europe and East Asia have long had separate and distinctive histories, during the Upper Paleolithic age the peoples of Eurasia followed the same way of life, hunted the same animals across the Eurasian steppe, and endured the same vicissitudes of climate. At the beginning of the Upper Paleolithic, there must have been some intertwining of the populations of west and east Eurasia since both were drawn from a common source, that of the first emigrants who reached India. Women who belong to the mitochondrial lineage known as X are witnesses of this distant bond. Some of X's daughters migrated northwest from India and are European; others traveled northeast, crossed Siberia and the land bridge to Alaska, and are now American Indians.

Only gradually did east and west diverge. The vastness of Eurasia inevitably pushed its Upper Paleolithic people into separate trajectories. This chapter follows the peoples of the west, as they took the slow and difficult steps toward settlement, and the peoples of the east as they domesticated the dog and discovered the Americas.

Upper Paleolithic Transitions

During the Pleistocene, much of northern Europe and northern Asia, or Siberia, was covered in glaciers, and the climate was much drier than now, with frigid deserts skirting the glaciers' southern edges. Because so much of the oceans' water was locked up in ice, sea level was more than 200 feet lower for much of the period and the map of the world, could people of the Upper Paleolithic have envisioned any such thing, was very different. Besides the since shrunken continents of Sunda and Sahul, a third land mass, one that is now totally submerged, lay between Siberia and Alaska. Beringia would serve as a broad land bridge to the Americas, but not immediately; for much of the time it was too dry to support vegetation and game, or the travelers who might depend on them.

With their talent for innovation, the people of the Upper Paleolithic quickly learned to live in the frozen north, drawn by the rivers of reindeer that flowed across the vast expanse of the Eurasian tundra. Over the millennia, as the climate and ecological conditions changed, they would switch to mammoth, then to ibex and red deer, and back to reindeer. One culture succeeded another in the archaeological record, but the foraging way of life remained a constant.

As the Pleistocene ice age drew toward its close, a dramatic change occurred in the world's climate. About 20,000 years ago, in an unparalleled catastrophe known as the Last Glacial Maximum, the glaciers came surging back for a last time, rendering most of northern Europe and Siberia uninhabitable. The world's population was probably between 1 and 10 million people at the time. A large proportion would have been affected by the sudden chill in climate. The people and animals of Eurasia survived only in southern refuges. Five thousand years later, the glaciers relaxed their hold on the Eurasian continent and retreated, allowing the survivors to move north once more.

Because Europe's archaeology, languages and genetics have so far received more attention than those of any other region of the world, Europe is the best theater in which to follow the history of human foraging societies.

On the basis of stone tools, archaeologists have distinguished a succession of European cultures in the Upper Paleolithic age. The earliest, the Aurignacian, lasted from 45,000 to 28,000 years ago. Sites with Aurignacian tools occur in France, Italy and much of eastern Europe, with an outlying province in the Levant. The Aurignacians were presumably accomplished fighters since it was they who steadily drove back and eventually exterminated the fierce Neanderthals. But their culture was not purely martial; it included the magnificent artists who decorated the Chauvet cave in the Ardèche Valley of France, the earliest known of the great painted caves of Europe.

The cave, according to the evidence of radiocarbon dates, was occupied at two periods, first from 32,000 to 30,000 years ago, and then again from 27,000 to 25,000 years ago.[129] Its walls are dominated by paintings of lions, mammoths and rhinoceroses, animals that were rarely hunted, according to archaeological evidence, as well as horses, reindeer, aurochsen, and an owl. The beauty and expressiveness of the paintings speak directly to contemporary observers. Yet despite the empathy they may arouse, the paintings' meaning, and the intent of their makers, is simply unknown. The natural assumption is that only people like ourselves could create such appealing works of art. But it is also possible that these are works of a savage intelligence that saw the world with the same visual system and a profoundly different mind.

Because of the cold climate that then prevailed, Europe and much of the

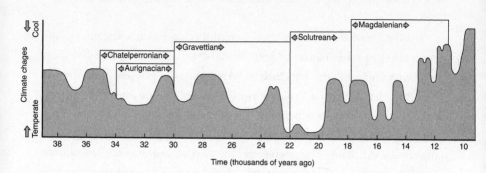

FIGURE 6.1. THE ROLLER COASTER OF CLIMATE CHANGE
IN THE UPPER PALEOLITHIC.

The Upper Paleolithic age in Europe was a time of sharp temperature changes, particularly during the Last Glacial Maximum, which lasted from about 20,000 to 15,000 years ago.

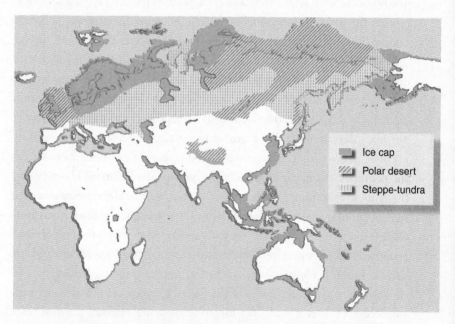

FIGURE 6.2. THE FORCED EVACUATION OF EUROPE AND ASIA.

During the Last Glacial Maximum northern and central Eurasia were covered with glaciers, bordered by steppe and tundra, and in both halves of the supercontinent the population would have been forced to migrate southward into warmer refuges.

Eurasian steppe were covered not in forest but in vast grasslands that supported abundant reindeer, woolly mammoths, bison and antelope. These animals provided ample subsistence for hunters, as well as valuable materials like hide, bone, ivory and antler.

The Aurignacian era came to an end, for unknown reasons, and its culture was replaced by that of the Gravettian, also defined by a distinctive set of stone tools. The Gravettian, which lasted from 28,000 to 21,000 years ago, stretched east into Russia, with southern provinces in Italy and astride the French-Spanish border. Gravettian people focused more on hunting mammoth than reindeer. They produced the well-known Venus figurines, with their dwarf heads, ample breasts and steatopygous buttocks, strangely reminiscent of the adaptation found among the San and the Andaman islanders. The figurines, recovered from sites stretching from France to Russia, clearly had some widely recognized importance in the Gravettian culture and were perhaps associated with a fertility cult. A less well known achievement is the invention of the bow—the earliest evidence of bows and arrows first appears at the end of the Gravettian period.[130]

The Gravettian culture occurred during a period of considerable cold during which much of the northern European plain was unoccupied. The era ended as the Last Glacial Maximum descended on the world. Its glaciers smothered Britain, Scandinavia and other northern latitudes, sending their occupants retreating to refuge areas in Spain, Italy and the Ukraine. Nothing is known about the collision of peoples that may have been set in train as the people of the north migrated down into the southerners' territory. But the worsening climate could have given an edge to the northerners who were adapted to the cold. The principal European culture during the post-Gravettian period is known as the Solutrean. It was centered in France and Spain and lasted from 21,000 to 16,500 years ago. Ibex, wild horse and red deer are the species whose bones are most common at Solutrean sites. The sites are more closely packed together, and some of the largest and thinnest stone tools look as if they were made for ceremonial rather than practical use. Archaeologists interpret these last two factors as a sign that people were living together in larger societies. This could have been a consequence of the fact that northwestern and central Europe had apparently been abandoned and the survivors were crowded into the southern refuges.

The Last Glacial Maximum lasted for some five thousand years. Then, as quickly as the glaciers had returned, they began their final withdrawal, yielding back the rich plains of northern Eurasia for occupation by animals and those who hunted them. From one of the refuge areas, the Périgord region of southwestern France, people spread out across the region that is now France and Germany, creating the Magdalenian culture which existed from 18,000 to 11,000 years ago. The Magdalenian tool kit, designed for reindeer hunting, is lightweight and portable. People crafted tools of particular precision and delicacy, such as bone harpoons with a row of barbs on each side. The practice of cave art continued at the Magdalenian sites of Lascaux, dated to 17,000 years ago, Niaux and Altamira.

Little is known about the lives of Upper Paleolithic hunter-gatherers or the reasons that led one culture to succeed another. For lack of contrary evidence, their social structure is generally assumed to have been egalitarian, without kings or leaders, as is that of contemporary hunter-gatherer societies.

Archaeologists are skilled at making inferences from the few shards of stone or bones they have to work with, but such evidence can lead only so far. Rarely can they identify the people who made the artifacts they study. Geneticists have begun to supply a new dimension to the archaeology by supplying biological information to match with the archaeologists' culture. Who were the Aurignacians or Gravettians? Amazingly, geneticists have been able to develop answers as to where they came from, and who their living descendants are.

The most comprehensive study so far of Europe's early population history has been carried out by Martin Richards of the University of Huddersfield in Britain. With colleagues in Europe and Israel, Richards has used an ingenious technique called founder analysis to date the arrival of successive waves of immigrants into Europe from 45,000 years ago to recent times.

Founder analysis depends on the idea that when people in region A send out colonists to region B, the colonists will start to clock up new mutations in their DNA that won't exist in the parent population back home in region A. So if the new mutations can be identified and counted, their number will yield an estimate of how long the colonists have lived in their new home in region B.

Richards has applied the founder analysis technique to mitochondrial DNA. As discussed above, mitochondrial DNA lineages have a distinctive geographical distribution because the mutations that initiate each branch of the genealogy occurred while people were moving into new territory across the world. The lineages denoted M and N were the only ones to come out of Africa and reach India. The daughter lineages of M and some of N populated all of the eastern Eurasian land mass; the rest of N populated western Eurasia.

N gave rise to a daughter lineage R, and the descendants of R, daughter lineages known as J, H, V, T, K and U, moved to occupy the Near East and Europe. Almost all Europeans belong to one or another of these six lineages or to a seventh, X, who is a direct, non-R daughter of N; hence the title of an engaging book by the population geneticist Bryan Sykes called *The Seven Daughters of Eve*. U, the most prolific daughter, had several sublineages of which, confusingly, K is one, and the others are labeled U1 through U6.

To reconstruct the population history of Europe, Richards and his colleagues started with the principal mitochondrial lineages in Europe, then looked for the present day descendants of their source populations in the Near East. They then compared regions of the mitochondrial DNA of the U5 cluster of lineages, say, in Europe with members of the U5 cluster in the Near East. After the two groups of U5 had parted ways, each would have continued to accumulate its own mutations. So it was easy to spot the new mutations in European U5—they were the ones that didn't also appear in Near Eastern U5.

Knowing the number of new mutations in European U5, and the general rate at which changes occur in mitochondrial DNA, the Richards team could then calculate how long U5 had been present in Europe. They performed the same exercise for the other main clusters of European lineages.[131]

They found that just eleven clusters, containing some 40 individual lineages, accounted for three quarters of the present day European population. The most ancient, the U5 cluster, had a time-in-Europe date of 50,000 years, give or take 5,000 years each way. This fits well with the archaeological date of 45,000 years for a site in Bulgaria that marks the earliest known presence of modern humans in Europe and hence denotes the start of the Upper Paleolithic age. And it suggests that the Aurignacians, the first people to enter Europe, belonged to the U5 mitochondrial lineage.

Richards could not tell how many people had entered Europe during this first entry of modern humans. But taking the number of lineages in the U5 cluster as a fraction of the whole, he calculated that about 7% of today's Europeans are descended from these first arrivals.

All but one of the other clusters arrived at various times in between, from 35,000 to 15,000 years ago. Altogether some 87% of Europeans are descended from people who arrived before the end of the Pleistocene ice age. Only 13% are descended from ancestors who came to Europe around 10,000 years ago, mostly in the form of the J cluster of mitochondrial lineages. These arrivistes would presumably represent the immigrants from the Near East who brought knowledge of agriculture into Europe and were the harbingers of the Neolithic age.

A major factor that shaped the present population of Europe were the glaciers of the Last Glacial Maximum which drove people back into the southern refuges of the Iberian peninsula. Europe was then repopulated by people spreading northeastward from these refuges as the glaciers retreated after 15,000 years ago.[132] The lineage clusters V and H, which had entered Europe earlier, were prominent in the reexpansion. Some 45 to 50% of most European populations belong to this cluster, and 60% of Basques do, as might be expected if the Basque region of southwestern France and northeastern Spain was the source of the recolonization.

The Richards team's reconstruction of the population history of Europe brought to light an unexpected fact: that most Europeans are descended from the first settlers who arrived during the Upper Paleolithic era. Only a minority arrived during the Neolithic age. This is the reverse of expectation; archaeologists had assumed that the people who introduced farming to Europe in the Neolithic age overwhelmed the earlier inhabitants with their larger populations. The findings suggest that the people of the Upper Paleolithic did not die out; they switched from foraging to settlement and adopted the new farming techniques.

Since men and women migrate together, studies of Y chromosome lineages should corroborate the conclusions drawn by the Richards group from mitochondrial DNA. To a large extent they do. A recent analysis by Ornella Semino of the University of Pavia and Peter Underhill of Stanford University has established that 95% of European men belong to just 10 lineages of the Y chromosome tree.[133] The researchers cannot find a Y chromosome lineage that matches up specifically with U5, the mitochondrial DNA cluster that signals the first arrivals in Europe 45,000 years ago. But they see evidence for lineages of men, all carrying a mutation known as M173, who arrived between 40,000 and 35,000 years ago and whom they consider the likely bearers of the Aurignacian culture.

A second migration of Y chromosome owners arrived in Europe from the Near East some 20,000 to 25,000 years ago. The mutation that defines these lineages is known as M170. These men seem to have been the bearers of the Gravettian culture that succeeded the Aurignacian, Semino and her colleagues say.

The work of the Richards and Semino teams lays the basis for what many hope will be a grand synthesis between genetics and archaeology. If the geneticists can firm up the dates of entry into Europe of the various mitochondrial DNA and Y chromosome lineages, archaeologists may be able to tie these population movements into the sequence of culturally distinct occupations they have defined for the Upper Paleolithic period. And if historical linguists should succeed in reconstructing a family tree of human languages, as discussed in chapter 10, it may even be possible to say what language was spoken by the people of these ancient lineages. Such a link can already be suggested in at least one instance: if people of the mitochondrial lineage J were indeed those who arrived 10,000 years ago bringing the agricultural techniques of the Neolithic, then they may have spoken the Indo-European tongue from which so many of today's European languages are descended.

The Upper Paleolithic in Eastern Asia

The population history of East Asia cannot yet be written in the same detail as that of Europe. Although the two halves of the Eurasian continent developed separately, there have clearly been links between them. One lies with the men who brought the Aurignacian culture to Europe. Their branch of the Y chromosome genealogy, defined by mutation M173, is a brother to the M3 lineage that is found in some Siberian populations and many American Indians; the two lineages presumably originated from the same source, perhaps in India. Upper Paleolithic sites similar to those in Europe and dating from 40,000 to 25,000 years ago are found across Siberia and around Lake Baikal.

The Siberians probably lived in much the same way as their European cousins, hunting the large herds of hoofed species that grazed the Eurasian steppe land. Like the Europeans, their millennia of foraging life were disrupted by the rigors of the Last Glacial Maximum.

Siberia may be something of a backwater in the contemporary world, but in the days of the Last Glacial Maximum its inhabitants accomplished two historic achievements. One was the domestication of the dog, the first species to be drawn into human service. The second, of lesser immediate importance, was the discovery and inhabitation of North and South America.

Dogs have lost their working status in most modern societies. But they spread like wildfire in the prehistoric world. They could be trained to help hunt other animals. They made good bed warmers during cold Siberian nights. They would have been a self-transporting source of meat in case of emergency. But probably none of these is the reason that dogs spread so quickly from one end of Eurasia to another.

In antithesis to the Sherlock Holmes tale that hinges on the dog that didn't bark in the night, a crucial problem of dog origins is why they do. Wolves almost never bark. Barking was probably a character that was selected by the dog's first domesticators. That suggests they weren't much interested in using dogs for hunting, where a bark is no asset. But if the first use of dogs was in sentry duty, to warn of strangers, intruders, and attackers creeping in for a dawn raid, then a fierce and furious bark would have made a dog an invaluable defense system.

Dogs may thus have played an important role in early human history, especially if they helped make possible the transition from foraging to settled societies. People who settled down in one place would have been under constant risk of attack. It is perhaps significant that the first settlements occurred at the same time as dogs were domesticated.

Dogs are wolves that have been genetically adapted to live with people. In biological relationships between two species, it is common for each to evolve in response to the other. Have people adapted so as to live with dogs? Communities that learned to make use of dogs as sentries may have gained a substantial advantage, especially in conditions of constant warfare, over those whose members did not learn how to establish rapport with dogs.

Another way in which dogs may have altered early human societies is by disrupting the foragers' taboo against private ownership of property.[134] Dogs don't belong to a community: they attach themselves to a master. Possibly they forced themselves into human societies as the first major item of ownership, paving the way for the concept of the property-based sedentary societies that were to follow.

Robert Wayne of the University of California, Los Angeles, has studied the mitochondrial DNA of dogs, wolves, coyotes and jackals, and shown that, as long supposed, dogs are almost certainly descended from wolves alone, even though all these canid species can interbreed.[135] But his estimated date for the origin of dogs as a separate population—135,000 years ago—seemed far too early to archaeologists. The oldest dog bone found so far, in Germany, is 14,000 years old, with other dogs 12,000 years old known from Israel.

A much more plausible date emerged from a subsequent survey by one of Wayne's colleagues, Peter Savolainen, now at the Royal Institute of Technology in Stockholm. Savolainen collected mitochondrial DNA, asking dog fanciers from all over the world to send him hairs from their breeds. With samples from 654 dogs from Europe, Asia, Africa and America, and 38 Old World wolves, he was able to pinpoint the likeliest region where dogs were domesticated as being somewhere in East Asia, even though the earliest known dog remains occur in the West. This is because there is more variability in the DNA of East Asian dogs than anywhere else in the world, and a rule of thumb in genetics is that the region of a species' greatest diversity is its place of origin.

Almost all the dogs in Savolainen's sample fell into three main clusters, suggesting either that they had been domesticated independently three times or that three related wolves from the same litter or pack had been domesticated at the same time and place. Savolainen favors the latter interpretation because it gives a more plausible date for the domestication event—15,000 years ago. (The alternative case of three separate domestications implies a date of 40,000 years ago. But an invention as useful as the dog would probably have spread like wildfire, and there is no evidence for dogs for another 26,000 years.)[136]

It's a considerable puzzle to understand how the process of domesticating wolves into dogs got under way. Some species can't be domesticated at all and with others, many generations of selective breeding are required to produce any results. The difficulties were demonstrated in a remarkable experiment by a Soviet scientist, Dmitri K. Belyaev, who set out to domesticate silver foxes. His theory was that all or most domestic animals had been derived from their wild forebears by the same straightforward criterion, that of tameability. The set of genes required to bring about this profound change of behavior in a wild animal, he believed, also induced the distinctive physical characteristics found in many species of domesticated animal. These include white patches on the pelt, curly hair, shorter tails and floppy ears.

Belyaev and his successor, Lyudmila N. Trut, selected silver fox puppies on the sole criterion of tameness, choosing only those least hostile to human contact as the parents of the next generation. After 40 years, 45,000 foxes, and 30 to 35 generations of breeding, Trut now has a population of 100 docile, eager-to-please silver foxes, many carrying the white patches that Belyaev had predicted.[137]

In the case of dogs, domestication has another ingredient besides tameability, which is the capacity to read human body language. Brian Hare of Harvard University has tested the ability of dogs, wolves and chimps to pick up on cues as to which container holds hidden food. The experimenter would give broad hints, such as tapping the right container, or staring at it. Chimpanzees have a lot more intellectual wattage than dogs, yet very few got the message because they paid no particular attention to what the experimenter was doing. Wolves too are very smart, but did not take the hint. But dogs, and even puppies, picked up instantly on the hint being conveyed.[138] Because even puppies have this ability, it is probably innate and would have been a behavior selected for in the domestication process, Hare concludes, though it may go along with tameability rather than being a separate behavior.

That still leaves open the question of what humans were hoping to achieve when they set about domesticating wolves, given that the eventual outcome could hardly be foreseen. Ray Coppinger, a dog behavior expert at Hampshire College, believes that people can take little credit for the process; it was wolves who domesticated themselves. Wolves are skillful hunters, but they also scavenge. They would have hung around campsites for scraps, and those that learned to be less afraid of people would have flourished, in his view.

"It was natural selection—the dogs did it, not people," he says. "The trouble with the theory that people domesticated dogs is that it requires thousands of dogs, just as Belyaev used thousands of foxes." From the semi-tame, camp-following wolves, he believes, people may have adopted some cubs into the household and found that they could be trained.[139]

Hunter-gathering peoples often bring baby wild animals back to camp and keep them as pets until they become unmanageable. James Serpell, an expert on dog behavior at the University of Pennsylvania, thinks this is a more likely basis for domestication than that people adopted wolves that had taken up life as scavengers. If the wolf was domesticated only once, from a group of related animals, there may have been some special feature of these wolves' behavior that made them easier to train, Serpell suggests.

However the bond between man and dog was first forged, it proved unbreakable. The Siberians who first ventured into North America via the lost continent of Beringia, the now sunken lands of the Bering strait between Siberia and Alaska, took their dogs with them. This is a surprising finding since researchers had assumed American Indians would have domesticated their own dogs from North American wolves. But Jennifer Leonard, of the University of California, Los Angeles, extracted ancient mitochondrial DNA from pre-Columbian dog cemeteries in Mexico, Peru and Bolivia. She found they matched the DNA of gray wolves from the Old World and not of wolves from the New World.

The DNA from the dog cemeteries clustered into five groups, suggesting that five different dogs, or sets of related dogs, entered the New World and were the founding mothers of all pre-Columbian dogs.[140] For unknown reasons, these pre-Columbian dog lineages have all disappeared. American Indians seem to have preferred the dogs brought in by Europeans. Breeds of dog that were developed in the New World, such as the Eskimo dog, the Mexican hairless dog and the Chesapeake Bay retriever, are all derived from European dogs.

The First Discovery of the Americas

Besides inventing the dog, the Upper Paleolithic people of East Asia made another historical contribution: they discovered and colonized the two major continents of North and South America. Genetic comparison of present

day Siberians and American Indians may at last be bringing some resolution
to two long running academic disputes about the settling of America. One
is linguistic, the other archaeological.

More than 600 languages are spoken by American Indians and they are
so different from each other that most linguists have regarded them as being
derived from several different language stocks. In many disciplines there are
lumpers, who see patterns and commonalities, and splitters, whose preference is to define differences. In historical linguistics, the splitters have the
majority. A leading, and generally lonely, lumper is the late Joseph Greenberg of Stanford, a maverick with a fundamentally different view on how to
establish the relationship among groups of languages.

In 1987 Greenberg caused more than usual distress among his fellow linguists when he announced his finding that all American languages fell into
just three major families. There was Eskimo-Aleut, a group of 10 languages
spoken by Eskimos and the inhabitants of the Aleutian islands off Alaska.
There was Na-Dene, a family of 32 languages spoken only in North America,
by the Apache, the Navajo and tribes in Canada and Alaska. And finally
there was Amerind, a group to which in his view all 583 other languages of
North and South America belonged.

Greenberg well understood that the specialists in various Indian languages would not embrace the idea that their beloved tongues were all splin-
ters off the same block. "I am therefore well aware that what is attempted in
this work runs against the current trends in Amerindian work and will be received in certain quarters with something akin to outrage," he wrote. "Given
the investment in time and energy that has led to results different from
mine, such a reaction is wholly understandable."[141] The implication that his
opponents' ardor was more substantial than their acumen reflects the general state of relations between Greenberg and his critics.

Greenberg broadened his linguistic classification into a sweeping and at-
tractively elegant hypothesis. He suggested that the three language groups
he had defined represented three separate waves of migration into the Americas from Siberia. There were independent reasons, he noted, from study of
teeth and of immunology, for assuming there had been three distinct waves
of immigrants. As might be expected, the three migrations are packed into
the Americas in order of arrival. The Amerind-speakers, who reach to the tip

of South America, were clearly the first. Greenberg suggested they entered America from Siberia about 12,000 years ago and he linked their arrival with the appearance of what archaeologists call the Clovis culture, the earliest indisputable evidence of human presence in the Americas. The Clovis people lived on the Great Plains and hunted mammoth and bison from 11,500 until about 11,000 years ago. At this date the mammoth and several other larger American species became extinct, a customary indicator of human arrival, although in this instance the ending of the Pleistocene ice age may also have been a factor.

Several thousand years after the Amerind migration came the Na-Dene speakers of northwest North America, Greenberg supposed, and last to arrive were the Eskimo-Aleut of the circum-Arctic.

The relatively recent date adduced by Greenberg for the first entry to the Americas supported his position that it should be possible still to see links between the various Amerind languages. But archaeologists have long been seriously divided over the question of first entry. No human remains older than those of the Clovis culture have yet been discovered, but there are hints of an earlier presence, notably at the Monte Verde site in southern Chile. One layer of apparent artifacts, mostly plant remains and wooden objects possibly associated with tents, has yielded radiocarbon dates of 12,500 years ago, while a deeper and more doubtful layer has produced dates of 33,000 years ago. After initial rejection and long debate, archaeologists have finally accepted the 12,500-year layer, though not the older stratum, as evidence of a pre-Clovis presence, according to a recent review.[142] This gives a date slightly before Clovis, but still leaves the impression, at least on archaeological grounds, that the two continents were empty of people prior to 14,000 years or so.

When the geneticists first arrived on this particular academic battleground, they generally favored the idea of a few migrations, though not necessarily just three. But mitochondrial DNA, the genetic element they were first able to analyze, pointed to much earlier dates for the colonization of the Americas, lending preliminary support to the archaeologists who favored seriously pre-Clovis dates of settlement such as 30,000 years ago.

There are five groups of mitochondrial DNA lineages in the Americas, the groups known as A, B, C and D, as well as the small lineage X, which has a special history. A, C and D are also found in northern latitudes of Asia and in northeastern Siberia but B has a different distribution, being found in southeastern Siberia. That led Douglas Wallace, of the University of California at Irvine, to suggest that the first entry into the Americas occurred 34,000 years ago and consisted of people carrying lineages A, C and D migrating from Beringia. There was then a second migration, 16,000 to 13,000 years ago, according to mitochondrial DNA evidence, that brought lineage groups B and X to the Americas. The Eskimos and Na-Dene speakers appeared on the scene sometime after 10,000 years ago, Wallace suggested.[143]

These dates fitted with those derived by several other research groups working on mitochondrial DNA, which ranged from 10,000 to 40,000 years ago. But it now seems that most of these dates, for reasons that are not wholly clear, may be far too old. A second group of geneticists has now entered the fray on the back of the Y chromosome, and they have made a strong case for much younger dates of entry, broadly coinciding with Greenberg's original thesis.

The Y chromosome is usually harder to date than mitochondrial DNA, but in the case of the Americas geneticists have been helped by finding a mutation on the Y that occurred just before the first entrants crossed into Beringia and the Americas. The mutation, known as M242, seems to have cropped up between 15,000 and 18,000 years ago, according to Mark Seielstad and colleagues.[144] It occurred just before another mutation on the Y chromosome, called M3, which is found almost exclusively in American Indians. So on this evidence the entry to the Americas could not have occurred earlier than 15,000 to 18,000 years ago.

The genetic, archaeological and linguistic data have been drawn together in what may prove to be a convincing synthesis by Andres Ruiz-Linares of University College, London. He and colleagues conducted a large survey of Y chromosome variations in Mongolians and in American Indians, mostly from South America. They conclude that two major waves of migration from Siberia account for the origin of American Indians.[145] Both waves originated ultimately from the southern latitudes of central Siberia. The first entered North America about 14,000 years ago and spread throughout both hemispheres. The second migration arrived later and remained restricted to North America. This migration may have originated from a region of Siberia occupied by the Kets, whose language Greenberg has suggested is related to Na-Dene.

Many of the tribes of South America show strong signs of genetic drift, an indication that their populations have bred in isolation for many thousands of years. Ruiz-Linares estimates from a DNA signature found in two tribes, the Ticuna of the upper Amazon and the Wayuu, on the north coast of Colombia, that they have been genetically isolated for some 7,000 to 8,000 years. The finding suggests that tribalization—the division into small, warring populations who each defended a home territory—started soon after the first migrants reached South America.

Early division and an ancient origin for South America's tribal populations would go a long way toward explaining why Amerindian languages have grown so different from one another in such a relatively short time. If this genetic interpretation is correct, it would explain why the linguistic splitters are right to point to the large differences between Amerindian languages but also why Greenberg was right in lumping all the languages together in a single family.

Should the new picture emerging from the Y chromosome be confirmed, it will lend support to Greenberg's idea of three waves of migration. The first arrivals would have crossed the Beringian land bridge after 14,000 years ago and before 11,000 years ago, when Beringia was submerged by rising sea level. They would have spread quite rapidly southward, perhaps by boat if the 12,500 years ago date for Monte Verde in southern Chile is correct. A second migration brought in the Na-Dene speakers, and sometime later the Eskimo-Aleuts arrived. The mitochondrial DNA lineages are now found in all three language groups, presumably because of subsequent mixing between them, but it seems that A, C and D predominated in the first migration, B and X in the second.

The mitochondrial lineage known as X has turned up in recent surveys among the Sioux, Navajo, Ojibwa and other tribes of North America. Its discovery at first caused considerable surprise because X is one of the founding lineages of Europe. The finding generated some colorful theories, such as that the women from the Vikings' unsuccessful Vinland colony in North America had been abducted by the skraelings, as the Vikings called their Indian assailants. But geneticists soon showed that the cluster of X lineages in North America was at least 12,000 years old,[146] and ancient mitochondrial DNA of the X lineage was discovered in bones some 1,200 years old from a site in Washington state, far distant from the Vinland colony and somewhat earlier.[147]

The explanation must be that X, a daughter lineage of N that reached India, participated in two major migrations. As already noted, some women of the X group of lineages moved westward into Europe while their sisters joined the expansion into the Central Asian steppes and Siberia. Many generations later, the descendants of these daughters of X were among the first discoverers of North America.

Adapting to Cold with Mitochondrial DNA

The people who reached the Americas from Siberia may have possessed a special quality. The bridge from Siberia to Alaska was not hard to spot— Beringia was a land mass the size of a continent before it sank. So why did only a handful of groups succeed in making their way across? An obvious possibility is that Siberia and Beringia were cold places in which not everyone could survive. It may have been no accident that people of the mitochondrial lineages A, B, C, D and X were the only ones to reach the Americas. According to a proposal by Wallace, these mitochondria may have conferred a special resistance to cold.

Human mitochondrial lineages, Wallace has pointed out, are geographically patterned not just by continent but also by latitude. The most ancient lineages, L1, L2 and L3, are specific to sub-Saharan Africa. It was only carriers of L3 who moved northward into northeast Africa, and only L3's daughter lineages, M and N, that left Africa to colonize temperate zones. Wallace wondered if that distribution might be not just a matter of chance, as generally assumed, but rather of natural selection. Mitochondria produce the body's energy and heat, and survival in cold and even temperate climates could depend a great deal on which lineage of mitochondria a person inherited.

Mitochondria release their output in the form either of heat or of an energy-carrying chemical known as ATP. The balance between heat and ATP production can vary, depending on DNA changes in the mitochondrial genes that operate the energy production system. People living in cold climates would be better off with mitochondria adapted to produce more heat and less chemical energy. If so, Wallace argued, their mitochondrial genes should show signs of having been under pressure from natural selection.* In testing mitochondrial DNAs from around the world, Wallace has found that some do indeed bear the marks of positive selection, particularly those of people who live in Siberia or whose ancestors did, such as most American Indians. The groups of mitochondrial lineages known as A, C, D and G are particularly common among arctic people; 75% of them belong to one of these four groups, but only 14% of Asians living in temperate zones do. Some European lineage groups, such as H, also show signs of adaptation to cold conditions.[148]

The adaptation of mitochondria to climate, Wallace believes, could explain why when you look at a map of the world the mitochondrial lineages seem to have a more limited distribution than the Y chromosome lineages. The Y chromosome carries rather few genes, most of them related to male fertility, and there is no reason to suppose it is affected by climate. This might explain why men have ranged farther afield than women, at least on a large scale. On a smaller geographical scale, the genetic evidence shows that women move farther than men, presumably reflecting the fact that most human groups are patrilocal and it is the women who move to other societies to find marriage partners.[149]

Adaptation to cold may have affected human populations in other important ways, particularly during the Last Glacial Maximum. In Europe, the ice sheets emptied all northern and central latitudes, which were repopulated many generations later by those who had survived in the southern refuges of Spain or the Ukraine. A similar phenomenon seems to have happened in the eastern half of Eurasia and may be responsible for one of the salient puzzles in human population history, the origin of the mongoloids.

Origin of the Mongoloid Peoples

"Mongoloid," a term from physical anthropology, refers to the skull shape typically found among East Asians and many American Indians. Skull shape figured prominently in racial theories of the nineteenth century, which

---

*The fingerprints of natural selection at work can be inferred by comparing silent mutations in DNA units (ones that don't change the design of a protein) with significant mutations (ones whose presence causes a different amino acid unit to be specified in the protein).

erroneously linked skull type with behaviors or abilities deemed characteristic of certain races. Modern craniometry, or skull measurement, is almost purely descriptive and has nothing to say about behavior. It depends on examining a large number of detailed anatomical features of the skull and making statistical correlations between them. Though these assessments are not easily translated into simple physical descriptions, contemporary East Asian skulls generally have fine features, broad head shape and flattened faces. Skulls vary from gracile to robust, terms used by physical anthropologists to denote the general thickness of the bone; mongoloid skulls are the most gracile in the human family.

Mongoloid skulls are also associated with a special kind of teeth. Many human groups, such as sub-Saharan Africans and Europeans, retain the generic, undifferentiated human teeth of the ancestral population. But people in southeast Asia, Polynesia, Australia, southern China and ancient Japan have developed a different dental complex known as sundadonty, after Sunda, the former continent that included Malaysia and much of Indonesia. A third category of teeth, itself derived from sundadonty, is sinodonty. Sinodonts include the people of northern China, modern Japan, and the native peoples of North and South America.[150] Mongoloids in general have both of the two derived types of teeth, with southern mongoloids being sundadont and northern mongoloids sinodont.

The puzzle is that mongoloid skull types, although now owned by the largest of all human racial groups, do not show up indisputably in the archaeological record until about 10,000 years ago. There were of course people in China before then, but those inhabitants possessed generic early modern human skulls. The mongoloid skull type is a very recent evolutionary development.

No one knows for sure what factors prompted the emergence of the mongoloid peoples from their predecessors, but two explanations have been suggested, both invoking the Last Glacial Maximum.

One is that the mongoloids emerged by genetic drift, the random fluctuation in gene frequencies that occurs between generations. Drift can lead to a single version of a gene becoming universal, or fixed, and all other versions being lost. Fixation of a gene depends on the size of the population, being faster in smaller populations, so anything that breaks a population into small, separately breeding communities will spur genetic drift and evolutionary change. The Last Glacial Maximum, by freezing the landscape in a patchy fashion, could well have fractured the habitat of the people living in the northern latitudes of East Asia into small populations subject to rapid drift. In one of these, presumably the most successful, the particular features of the mongoloid skull would have evolved by chance alone (since drift is a random process) and that group went on to dominate East Asia.

Another proposal is that the mongoloid skull type arose from natural selection. Biologists have long speculated that mongoloid features are an adaptation to cold. An extra layer of fat in the eyelid (the epicanthic fold) gives the eye more insulation. Pale skin lets in more sunlight, which the body requires for synthesis of vitamin D. A stockier body reduces heat loss. It's a plausible guess that genes favoring such features would have grown more frequent during the 5,000 or so years of the Last Glacial Maximum.

Drift and selection can of course act together. "It is possible that with the onset of glacial conditions the widespread population of eastern Asia contracted its range in its northern latitudes, resulting in a number of temporarily isolated groups," writes the physical anthropologist Marta Mirazón Lahr. "Under strong environmental pressure, morphological change could have become rapidly fixed in a population of small size."[151] Or, in less technical language, new versions of genes that favored the mongoloid physical appearance could have become universal in one of these groups through the selective pressure of the cold climate. East Asians seem to have evolved light skin independently of Europeans.[152] They also have a gene that leads to a dry form of earwax and less sweating.[153] When the glaciers retreated 15,000 years ago, the mongoloids would have expanded northward, like their counterparts did in Europe.

The first modern humans who migrated out of Africa almost certainly had dark skin, as do their descendants in Australia and the relict populations who still survive at points in between. Given that early modern human skulls are all much the same, it's possible that for many thousands of years all modern humans outside Africa, as well as those inside, had black skin. But at some stage, populations in both the western and eastern halves of Eurasia must have evolved into, or been replaced by, people with lighter skin. When that happened is at present a matter of speculation. But one point at which replacement could have occurred is during the Late Glacial Maximum. The populations living in northern latitudes had perhaps developed lighter skin, either for reasons of vitamin D synthesis or through sexual selection, by 20,000 years ago. When the glaciers returned, the cold-adapted northerners would have moved slowly south, along with the frigid climates to which they, but not their southern neighbors, were adapted. The freezing temperatures could have given them an edge in displacing their darker-skinned cousins in southern latitudes. Later, after the glaciers' retreat, the populations that expanded from their southern refuges, both in Europe and East Asia, would have been the descendants of the light-skinned northerners.

This might explain why the regional variations in skull type that characterize caucasoid peoples (those of western Eurasia: India, Europe and the Near East) and mongoloids (peoples of East Asia) do not become evident until the Holocene, the warm period that succeeded the great Pleistocene ice age 10,000 years ago. "Most early modern skulls do not exhibit unequivocal characteristics of any present-day race," writes the paleoanthropologist Richard Klein, "and it seems increasingly likely that the modern races formed mainly in the Holocene, after 12–10 ky [thousand years] ago. This is perhaps particularly clear for eastern Asia (the present-day hearth of the 'Mongoloids'), but it also applies to Europe (the homeland of the 'Caucasoids')."[154]

With the end of the Last Glacial Maximum, the dominance of the hunter-gathering way of life, the only kind of existence humans had ever known, also began at last to wane. It was in the Near East that the first sustained experiments in settled living were about to begin.

7

SETTLEMENT

Man accumulates property and bequeaths it to his children, so that the children of the rich have an advantage over the poor in the race for success, independently of bodily or mental superiority. . . . But the inheritance of property by itself is very far from an evil; for without the accumulation of capital the arts could not progress; and it is chiefly through their power that the civilised races have extended, and are now everywhere extending their range, so as to take the place of the lower races. Nor does the moderate accumulation of wealth interfere with the process of selection. When a poor man becomes moderately rich, his children enter trades or professions in which there is struggle enough, so that the able in body and mind succeed best.

CHARLES DARWIN, *THE DESCENT OF MAN*

THE LAST GLACIAL MAXIMUM preceded the emergence not only of people who looked somewhat different from each other but, far more significantly, of people who behaved differently from all their predecessors. In the southern borders of the western half of Eurasia, around the eastern shores of the Mediterranean, a new kind of human society evolved, one in which hunters and gatherers at last developed the behaviors necessary for living in settled communities.

The Pleistocene did not depart quietly but in a roller coaster of climatic swings. After the Last Glacial Maximum, of 20,000 to 15,000 years ago, came a warming period known as the Bølling-Allerød Interstadial, during which plants, animals and people were able to move northward again. But the Bølling-Allerød warming, which lasted from 15,000 to 12,500 years ago, was a false dawn. A second cold period, particularly challenging because it began so abruptly, established its grip on Eurasia. Within a decade, it had sent temperatures plummeting back to almost glacial levels and soon had converted to tundra the vast forests of northern Europe. This deadly cold snap is known as the Younger Dryas, after a dwarf yellow rose, *Dryas oc-topetala*, that grew amid the tundra.

The Younger Dryas lasted for 1,300 years and ended as suddenly as it began, also in a decade or so, according to the cores drilled from through the Greenland ice cap that serve as an archive of global climate. By 11,500 years ago the world was launched on the Holocene, the inter-ice age period that still prevails.

These wrenching climatic and territorial changes would have posed severe tests to human survival, doubtless forcing people to resort to many new expedients even in the warmer southern latitudes. The precise chain of cause and effect, if any, remains a mystery. All that can be said for now is that in the Near East, as the Last Glacial Maximum ended, a new kind of human society began to emerge, one based not on the narrow ambit of the forager's life but on settling down in one place.

Settling down, or sedentism, as archaeologists say, may sound so simple and obvious, but for foragers it was not nearly so clear a choice. Sedentism tied people to a single exposed site, increasing vulnerability to raiders. Sedentism attracted noxious vermin and disease. Sedentism required new ways of thought, new social relationships and a new kind of social organization, one in which people had to trade their prized freedom and equality for hierarchy, officials and chiefs and other encumbrances.

Archaeologists have little hesitation in describing the transition to sedentism as a revolution, comparable to the one that defines the beginning of the Upper Paleolithic 50,000 years ago when behaviorally modern humans emerged from their anatomically modern forebears. Ofer Bar-Yosef of Harvard University refers to these transitions as "two major revolutions in the history of humankind."[155]

Hunter-gatherers own almost no personal property and, without differences of wealth, everyone is more or less equal. The first settled communities show evidence of a quite different social order. Houses and storage facilities seem to have been privately owned. With personal property allowed, some people quickly acquired more of it than others, along with greater status. The old egalitarianism disappeared and in its place there emerged a hierarchical society, with chiefs and commoners, rich families and poor, specializations of labor, and the beginnings of formal religion in the form of an ancestor cult.

"Daily life in a village that is larger than a forager's band heralds the restructuring of the social organization, as it imposes more limits on the individual as well as on entire households," writes Bar-Yosef. "To ensure the long-term predictability of habitable conditions in a village, members accept certain rules of conduct that include, among other things, the role of leaders or headmen (possibly the richest members of the community), active or passive participation in ceremonies (conducted publicly in an open space) and the like."[156]

Sedentism must also have included a response to the most pressing of human social needs, defense against other human groups. For hunter-gatherers, the essence of security is mobility. For the first settlers, defense must have rested on some other basis, which was presumably that of population size. Because the settlers had learned to live together in larger groups, they would have outnumbered the attackers. With greater manpower than the usual foraging group, together with fortifications and perhaps the guard dogs that first became available 15,000 years ago, settlers would have been able to even the odds against the raiding parties after their food and women.

This new form of social organization preceded and perhaps prompted such innovations as the cultivation of wild cereals, and the penning and herding of wild animals like sheep and goats. These steps led in turn, perhaps more by accident than design, to the domestication of plants and animals and to the beginnings of agriculture. Settled life and the new hierarchical form of society paved the way for complex societies, cities, civilization and, in rudimentary form, the institutions of today's urban life. Almost all subsequent human history and development seems in one sense a consequence of the pivotal transition from the foraging lifestyle to a settled, structured society.

The innovations of settled life and agriculture started to spread through Europe 10,000 years ago, a date that marks the beginning of the Neolithic age. Because the two inventions became so visible in the Neolithic, archaeologists long assumed that the improving climate made agriculture possible, which in turn opened the gateway to settled living. But in part because of improved dating techniques, they have come to see that the reverse is true: it was not agriculture that led to settlement, but rather sedentary life came first, well before the Neolithic age began, and agriculture followed in its train.

"Until recently, the beginning of the Neolithic was thought to occur with the inception of village farming," write the archaeologists Peter Akkermans, of Leiden University in Holland, and Glenn Schwartz of Johns Hopkins University. "We are now aware, however, that sedentary village life began several millennia before the end of the late glacial period, and the full-scale adoption of agriculture and stock rearing occurred much later, in the late ninth and eighth millennia BC. It is now evident that agriculture was not a necessary prerequisite for sedentary life, nor were sedentary settlers always farmers."[157]

Some signs of sedentary life can be seen as early as the Gravettian mammoth bone houses of 18,000 years ago, and it may be that sedentary systems were attempted when people came across an abundant food source, such as hazelnuts or salmon, together with a method of storing it. But these early instances of settlement were sporadic and may not have required any deep behavioral changes. True sedentism did not catch on as a permanent way of life until toward the end of the Upper Paleolithic. The first clear evidence of a successful and long term settled community comes from people called the Natufians, who lived in the Near East from about 15,000 to 11,500 years ago. They occupied lands on the eastern side of the Mediterranean, in the region that is now Israel, Jordan and Syria. The early Natufians gathered the wild emmer wheat and barley that grew there. They made stone sickles to cut the cereal grasses, and the sickles bear signs of the characteristic polish caused by the silica in cereal stalks.[158]

Bar-Yosef suggests that the Natufians may have started to cultivate these wild cereals, including einkorn and emmer wheats, rice and barley, during the Younger Dryas when the natural yields of these cereal grasses would have been reduced. There is little evidence on the point, and in any event the Natufians did not develop the domesticated forms of the cereals. But in gathering, preparing and storing these grains, they were laying the technical basis for their successors to do so.

It is of interest that the Natufians, as the earliest known settled people, were no strangers to war or to religion, two characteristic human activities that shaped societies before and since. The Natufians have consistently been portrayed as peaceful but closer examination of remains from one site has recently shown evidence of violent conflict between Natufian groups.[159]

Natufian society is interesting for its burial practices, which indicate the

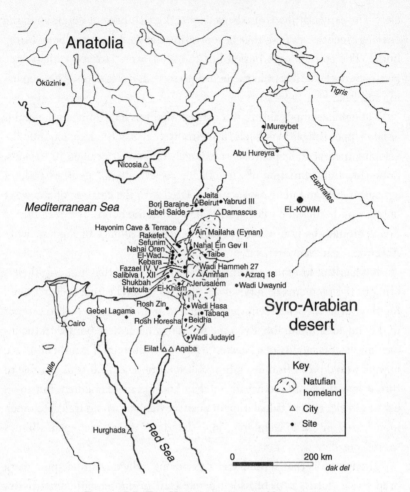

**FIGURE 7.1. THE HOMELAND OF THE NATUFIANS,
THE FIRST FORAGERS TO SETTLE.**

The Natufians built settlements on the east coast of the Mediterranean some 15,000 years ago. Later, they began to harvest wild stands of wheat and barley, laying the basis for others to develop domesticated forms of those cereals several thousand years after them.

emergence both of social inequality and of a disconcertingly intimate form of ancestor worship. Some 10% of early Natufian burials include decorations of marine shells and pendants made of animal teeth, suggesting the presence of a richer elite. In the later Natufian period, as the rigors of the Younger Dryas began, the society was forced to become more mobile, and their mortuary practices reflect a shift back toward a more egalitarian society. The early Natufians also began a practice that became common in the ensuing Neolithic period, that of separating the skull from the body before burial. The corpses were buried but the skulls were covered with plaster, given new faces, and kept in the houses to serve as a close bond between living and dead.[160]

Though it is impossible to reconstruct what was happening in the minds of late Upper Paleolithic people, it seems likely that settled life required developing mental concepts that were largely unfamiliar or alien to foragers. "The slow transformation of the foraging society into a Neolithic world of agriculturalists and herdsmen was associated with the creation of a new set of social and economic values centering around the house, the dead buried in and around the house, and the production and storage of staples," write Akkermans and Schwartz.

It is hard not to admire the fortitude and intelligence that hunter-gatherers bring to the problems of survival. But the set of intellectual skills required for survival in the wild seem quite different from those needed to prosper in the jungle of urban life. Even if a hunter-gatherer were born with the innate intellectual ability of a Newton, Darwin or Einstein, it is difficult to see how he would profit from his gift or, in evolution's cold calculus, be able to turn it into the reproductive advantage of raising more children. But in an urban setting, gifts of calculation or abstract thought would translate much more easily into extra children, and the genes underlying such abilities would spread.

The reason is that settled societies permit individuals to acquire extra property or status, both of which barely exist in hunter-gatherer societies and are in any case frowned on by their egalitarian ethos. Property, in turn, is a way of securing survival for oneself and one's family. For long periods of human history possession of excess property probably helped people raise more children, even though a direct relationship between wealth and progeny is not so evident in modern societies. Settlement, in other words, would have created a quite novel environment, to which people probably adapted by developing a different set of behaviors, including a range of intellectual skills for which there was no demand in hunter-gatherer societies.

Property, value, number, weight, measurement, quantification, commodity, money, capital, economy—these concepts, however natural to the modern mind, would rarely have come into play in the life of mobile foragers. Could it be that the modern mind, the one capable of abstract thought, symbolic notation and writing, is indeed a quite recent development? Perhaps the process by which the modern mind emerged "has to be regarded as a more gradual one, operating in several phases and stages, and perhaps independently in different parts of the world," writes the Cambridge archaeologist Colin Renfrew.[161]

That, in his view, might explain why human societies apparently accomplished so little for so long. "If human societies of the early Upper Paleolithic had this new capacity for innovation and creativity which notionally accompanies our species, why do we not hear more about them?" he asks. There is a 45,000-year delay between the time of the ancestral human population and the first great urban civilizations, such as those of Babylon, Egypt, the Harappan cultures of India and the Shang period of China. If "behaviorally modern" humans evolved 50,000 years ago, why did it take so long for this modernity to be put into practice? Renfrew calls this gap the "sapient paradox."

One possibility is that some evolutionary adaptation had first to occur in human social behavior. That would explain why it took so many generations for people to settle down. The adaptation, probably mediated by a suite of genetic changes, would have been new behaviors, perhaps ones that made people readier to live together in larger groups, to coexist without constant fighting and to accept the imposition of chieftains and hierarchy. This first change, of lesser aggressiveness, would have created the novel environment of a settled society, which in turn prompted a sequence of further adaptations, including perhaps the different set of intellectual capacities that is rewarded by the institution of property.

A striking change that preceded settlement is a worldwide thinning or gracilization of the human skull. This change, discussed further in the next chapter, was probably accompanied by a taming or greater sociability, doubtless a necessary step toward settling down in larger groups.

If such a change occurred, it evidently evolved independently in different regions of the world, just as have other human adaptations like pygmy stature and lactose tolerance. Direct evidence for such a change may emerge in time from the human genome once the genes that influence human social organization are identified.

Once people were settled, many new opportunities for human innovation were opened up in technology, trade, warfare and political organization.

A salient new technology was that of agriculture, which was invented before the end of the Pleistocene ice age and took off as soon as the climate started to warm up in the Neolithic. The reason for agriculture's rapid spread, archaeologists believe, was that societies of the Near East had preadapted to it, primarily by sedentism but also with efforts to intensify production by seeding wild grasses.[162] Many previous theories about the invention of agriculture have invoked external forces that allegedly pushed a passive human society into taking up cultivation. None is well supported. One thesis holds that population pressure drove people to agriculture. But the archaeological evidence is that human populations grew after the advent of agriculture, not before it. Another proposal is that the warming of the climate after the end of the Pleistocene ice age was the driving force. But climate improvement was much the same everywhere, yet agriculture emerges at very different times in different regions of the world.

"It is important to realize," write Akkermans and Schwartz, "that farming was neither the production of food according to an economic rationale nor an inevitability imposed on early Neolithic communities by large-scale events beyond their control. Instead, the adoption of agriculture was part of the profound transformation of the entire forager society and an adjustment to a wholly different set of societal values and meanings."[163] Sustenance is not the only reason for agriculture. One advantage enjoyed by settled societies, and denied to foragers, is the ability to generate and store surpluses. Surpluses form the basis for trade. They can be exchanged for things considerably more vital than extra food, like weapons, or alliances, or prestige.

Settlement and Domestication

By the end of the Pleistocene ice age 10,000 years ago, the second human revolution was well in place, that of reengineering the mobile, kin-based, foraging band into a settled society, bindable by ties of altruism and religion into larger groups. Societies of the Near East were the first to take this crucial step, one that enabled human inventiveness to thrive in a new setting. Specialization of roles may have occurred for the first time, which would have led to increased productivity. Productivity creates surpluses, and surpluses of one commodity can be traded for another with a neighboring group.

Settlement, specialization, property, surplus, trade—these are the sinews of economic activity, setting humans at long last on a separate path from living off nature's bounty like all other species.

Late Pleistocene peoples like the Natufians developed the technology of threshing and milling wild grains they had collected. They also began to cultivate wild grains, perhaps when the cold snap of the Younger Dryas shrank the natural expanses on which settled communities had become dependent.

It would only have been a short step from cultivating natural wild grasses to selecting specific types. The step may have taken place unwittingly. Einkorn wheat, emmer wheat and barley, three wild cereals that grow in the region of the Fertile Crescent, all have the property of shedding each kernel from an ear as it ripens. The domesticated varieties, on the other hand, keep all the kernels attached so all can be harvested together. If people harvested the wild varieties by knocking the sheddable ears off into baskets, any rare nonshedding mutant would be left to the end of the harvest. These would have served as the seed stock for the next generation, and the unconscious selection for nonshedding varieties would quickly have driven up the frequency of the nonshedding gene.

Unconscious selection may also have eliminated another undesirable feature of wild cereal grasses—their ability to inhibit their germination so as to avoid the trap of developing in a drought year.[164] Seeds that decided not to germinate would have been automatically eliminated in favor of mutants that did so in all weathers.

The transformation of cultivated wild cereals into their domestic forms could have happened very quickly, in as little as 20 to 30 years. That and other genetic considerations have been taken to mean that domestication of wheat was easy and might have happened several times independently.[165] But a genetic family tree drawn up for domesticated and wild varieties of einkorn wheat shows that the domesticated varieties all cluster on one branch, indicating a single domestication. The same is true for barley.[166]

Archaeologists have not so far found any single site where they can trace the progression from the wild form of a cereal to its domesticated versions. But genetics has provided an unexpected helping hand in the case of einkorn wheat. Francesco Salamini, of the Max Planck Institute for Plant Breeding Research in Cologne, Germany, with colleagues in Norway and Italy, analyzed nearly 1,400 strains of wild einkorn wheat from the Near East.

Those with a genetic structure closest to the domesticated strains came from the Karacadağ mountains of southeastern Turkey. The region is close to sites in northern Syria, like Abu Hureyra, where domesticated einkorn is known to have been grown some 8,500 years ago. The researchers conclude that "the Karacadağ mountains are very probably the site of einkorn domestication," a claim disputed by some but endorsed by Daniel Zohary, a leading expert on plant domestication.[167]

Einkorn was apparently the first wild cereal to have been domesticated. It was cultivated some 12,500 years ago and the first possible domesticated forms occurred 10,500 years ago; domesticated einkorn becomes abundant in the western half of the Fertile Crescent (from southeastern Turkey down the east Mediterranean coast) from 9,500 years ago. Domesticated emmer wheat, which is easier to harvest, is found at Abu Hureyra from 10,400 years ago. (Einkorn wheat mostly ceased to be planted in the Bronze Age; emmer is still grown in Ethiopia. Modern wheats stem from an accidental cross between a domesticated variety of emmer wheat and a wild grass known as *Aegilops squarrosa* or *tauschii*. The hybridization is thought to have occurred in the region of northern Iran some 7,000 years ago.) Rye and barley were two other wild cereals domesticated before 10,000 years ago in the Fertile Crescent.[168]

After the dog, the first animals to have been domesticated were sheep and goats, probably between 10,000 and 9,500 years ago. Cattle were domesticated from the aurochs at about the same time, and the pig from wild boar. The aurochs ranged widely across Europe as well as the Near East, but a comparison of British aurochsen (based on mitochondrial DNA extracted from fossil bones) with modern cattle shows that Europe's cattle too were domesticated in the Near East.[169] It may be that these animal species, like the wild cereals, were domesticated unconsciously, in a process that started with wild herds being penned and the tamer animals picked as parents of the next generation. This assumes that people of 10,000 years ago were not aiming at domestication because they had no idea it could be achieved. On the other hand, they had the dog as an example, and a growing number of instances of their own success.

The horse appears to have been domesticated much later and outside the Near East, probably on the Eurasian steppes. Wild and domesticated horse bones are hard to tell apart, but horse remains with possible bit wear on the teeth occur in archaeological sites of the Ukraine and Kazakhstan, starting from 6,000 years ago. Unlike other animal species so far studied, which appear to have been domesticated only once or twice, horses seem to have been domesticated on many separate occasions, according to a study based on mitochondrial DNA.[170] Possibly it was the technology for capturing, taming and rearing wild horses that spread from one society to another, rather than a strain of domesticated animals. If so, this would suggest that horses were of such high value, perhaps for military purposes, that people rushed to domesticate their own rather than waiting to acquire a breeding pair.[171]

The people of the Near East, having developed suites of domesticated plant and animal species, expanded their farming activities north and west into Europe. Archaeologists have generally assumed that these farmers could support more people and that their populations must have crowded out the original inhabitants of Europe who had entered as foragers during Upper Paleolithic times. But the founder analysis undertaken by Richards, as mentioned in the previous chapter, shows that only a small percentage of today's Europeans are descended from those who entered from the Near East in Neolithic times.

Presumably a few farmers from the Near East entered Europe, and perhaps the original inhabitants started to imitate their success, by settling down and adopting the new technology. Or the new farming groups, if composed largely of men in search of new land, may simply have captured women from the indigenous groups. The farmers' Neolithic genes would have become more diluted, generation by generation, as they and their new culture pushed farther into Europe.[172]

Whatever the mechanism of spread, only 4 of the 10 principal Y chromosome lineages found in today's Europeans arrived during Neolithic times. These 4 lineages, according to Semino and Underhill, account for 22% of European Y chromosomes, a reasonable match with mitochondrial DNA data suggesting that 13% of Europeans have Neolithic heritage.[173]

It is only a coincidence of timing that associates these Y chromosomes with the Neolithic, and, given the approximate nature of dates derived from genetics, it would be reassuring to have some more direct link. One has emerged from the painted pottery and figurines associated with Neolithic sites. The pottery, known as LBK from the German words for "linear band ceramics," was made in the Near East, the home of the Neolithic revolution, as well as in Greece, the Balkans and southern Italy. Two Stanford University researchers, Roy King and Peter Underhill, matched the geographical distribution of LBK pottery and figurines with that of the four Y chromosome lineages that entered Europe at the beginning of the Neolithic age. They found that one lineage in particular, marked by the mutation known as M172, was found in almost exactly the same locations as the LBK culture.[174] The present day male population with the highest known frequency of M172 happens to live in Konya, a city near the southern coast of Turkey and some 60 miles from the well known Neolithic site of Çatal Höyük. No less than 40% of men in Konya carry M172 on their Y chromosomes.

The finding supports the idea that Neolithic farmers from the region of Çatal Höyük pushed into Europe, gradually mixing with the local population. Their farming techniques and pottery making became universal, even though their genes did not. The intriguing question of whether they introduced the Indo-European languages into Europe is addressed in chapter 10.

The Interaction of Genes and Culture:
Lactose Tolerance

While people were shaping the genetics of domesticated plants and animals by altering various features of their environment, a curious thing was happening to people themselves. Their genetics too were changing as they adapted to the new environment of settled societies.

The warriors and mighty hunters who left the most children in hunter-gatherer societies may have lost their advantage in settled societies. The ability to support many children would have passed to those who excelled at the new occupations of farmer, priest, clerk or administrator. After many generations, and maybe not so many if the selection pressure was intense, people in settled communities may have developed a distinct suite of behaviors that set them apart from their hunter-gatherer forebears.

This conjecture cannot yet be addressed, because the genes that underlie human behavior are still for the most part unknown. But the ease with which the human genome responds to cultural changes in society has come to light from a physiological adaptation, the unusual ability to continue to digest lactose in adulthood, otherwise known as lactose tolerance.[175]

Though cattle were first domesticated in the Near East, Europe became a center of cattle breeding during one of its first farming cultures, known from its pottery as the Funnel Beaker culture. The culture, which lasted from 6,000 to 5,000 years ago, was located in north-central Europe in the region that now includes the Netherlands, northern Germany, Denmark and southern Norway. It has left a lasting mark on the genetics of both the cattle and human populations of the region.

A team of European researchers led by Albano Beja-Pereira recently studied genes that encode the 6 most important milk proteins in 70 breeds of European cattle. From samples taken from 20,000 cattle, they drew up a map showing the degree of genetic diversity in the cattle genes. The greatest diversity—usually the sign of a species' original homeland—coincided closely with the territory archaeologists have defined for the Funnel Beaker culture.

The researchers then performed the same mapping exercise for the human genetic trait known as lactose tolerance, the ability to digest lactose in adulthood. They found that the highest percentage of people with lactose tolerance occurred among populations in a region that substantially overlapped with the ancient territory of the Funnel Beaker culture. The frequency of lactose tolerance dropped off progressively with distance among populations outside the core area.[176]

This finding is remarkable because it shows a human population evolving, in recent times, in response to change created by human culture. Lactose is a special sugar that accounts for most of the caloric content of mother's milk. The gene for lactase, the enzyme that digests lactose, is switched on just before birth and, in most people, switched off after weaning. Because lactose does not occur naturally in most people's diet, it would be a waste of the body's resources to continue making the lactase enzyme. But in people of mostly northern European extraction, and to some extent in African and Bedouin tribes that drink raw milk, the lactase gene remains switched on to early adulthood or throughout life. Among these milk drinkers, the ability to digest the lactose in cow's, sheep's or goat's milk evidently conferred so great a benefit that the genetic mutation conferring the ability became widespread.

Geneticists are still trying to define the exact genetic change that causes the lactase gene to stay active after weaning. The DNA sequence of the lactase gene itself is identical in both lactose tolerant and intolerant people. The difference must lie in some nearby region of DNA that controls the activation of the lactase gene, such as the two mutations recently discovered by Leena Peltonen of the University of Helsinki.[177]

What is certain is that lactose tolerant Europeans have inherited unchanged from a common ancestor a huge block of DNA that includes the lactase gene, its neighboring gene and much else. The size of the block is a sign of recent evolutionary change. Big blocks of unchanged DNA are very rare because at each generation pairs of chromosomes swap sections of DNA so as to create individuals with novel combinations of genes. As is easy to envisage, the blocks of original DNA that a chromosome may start off with will get smaller and smaller at each generation as the swapping process whittles them down. So a large block of DNA shared by lots of people is a sign of recent selection. Large blocks are created when some must-have mutation occurs that is greatly favored by natural selection. Nature cannot pick out a specific mutation or gene; it can only favor individuals who have inherited the large block of DNA within which the advantageous gene occurs.

Besides indicating the presence of a gene under natural selection, a block of DNA can also be used to date the time the gene started to be selected, since the larger it is, the more recent the selection. Joel Hirschhorn of the Harvard Medical School has found that the block containing the lactase gene in lactose tolerant Europeans extends for about 1 million DNA units. He and colleagues believe that this is a sign of strong positive selection, and that the block started to become widespread sometime between 2,000 and 21,000 years ago.[178] This date fits with that of the Funnel Beaker culture.

Lactose tolerance occurs in a high percentage of many northern Europeans who live in the former region of the Funnel Beaker culture—in 100% of Dutch people, according to one survey, and 99% of Swedes. The condition also occurs in many other populations, though at generally much lower rates. In Africa, tribes who keep cattle, sheep or goats have higher rates of lactose tolerance than nonpastoralists. Lactose tolerance in some African groups includes as much as 25% of the population. It is presumably less common in these African groups than in northern Europeans because pastoralism got started later in Africa and natural selection has had less time to raise the frequency of the gene.

Lactose tolerance seems to have a different genetic basis in Africa because the DNA differences found by Peltonen and colleagues to be diagnostic of lactose tolerance in Europeans are largely absent from Africa.[179]

The phenomenon of lactose tolerance draws attention to three aspects of human evolution. First, it confirms that evolution didn't stop 50,000 years ago, when modern humans left Africa, as is often assumed, but has continued to reshape the human genome.

Second, it shows the human genome is likely to have responded independently in different populations to the same stimulus, a process known as convergent evolution. Lactose tolerance has arisen independently in northern Europeans and in several African populations. Many other human attributes that have evolved since the African diaspora may also have taken place independently in different populations, such as the probable cognitive advances discussed in chapter 5.

Third, the lactose tolerance phenomenon establishes that genes respond to cultural changes. This is not so surprising because culture is a major part of the human environment, and genomes are mechanisms for responding to the environment. But a feedback of culture on genes is rarely considered by social scientists, many of whom assume that human evolution ended for all practical purposes when cultural development began. The case of lactose tolerance shows that any long lasting human cultural behavior, such as drinking raw milk, can cause genetic changes if there is a way for the genome to respond to it.

Looking back on the years between 50,000 and 5,000 years ago, from the time of the ancestral human population to that of the Funnel Beaker people and their contemporaries, it is clear that wrenching changes in the human environment took place during this period, particularly in the social environment. Hunter-gatherers learned to settle down and cooperate in larger groups with people to whom they had no kin relationship. People who had been egalitarian and generalist joined hierarchical societies in which occupations were increasingly specialized. All these changes probably induced different behaviors, some of them maybe mediated through evolutionary changes to the human genome.

Human nature, in other words, has probably changed significantly in the last 50,000 years. It cannot have changed profoundly, because the principal lineaments of human nature are the same in societies around the world, suggesting that all are inherited from a single source. But any characteristic with a genetic basis can vary, and is very likely to do so, because few genes remain constant for long periods of time. The question of human nature and its evolution is the subject to be considered next.

8

SOCIALITY

Every one will admit that man is a social being. We see this in his dis-
like of solitude, and in his wish for society beyond that of his own fam-
ily. Solitary confinement is one of the severest punishments which can
be inflicted. . . . It is no argument against savage man being a social
animal, that the tribes inhabiting adjacent districts are almost always
at war with each other; for the social instincts never extend to all the
individuals of the same species. Judging from the analogy of the ma-
jority of the Quadrumana, it is probable that the early ape-like pro-
genitors of man were likewise social; but this is not of much
importance for us. Although man, as he now exists, has few special in-
stincts, having lost any which his early progenitors may have pos-
sessed, this is no reason why he should not have retained from an
extremely remote period some degree of instinctive love and sympathy
for his fellows. . . . As man is a social animal, it is almost certain that
he would inherit a tendency to be faithful to his comrades, and obedi-
ent to the leader of his tribe; for these qualities are common to most
social animals. He would consequently possess some capacity for self-
command. He would from an inherited tendency be willing to defend,
in concert with others, his fellow-men; and would be ready to aid them
in any way, which did not too greatly interfere with his own welfare or
his own strong desires.

CHARLES DARWIN, THE DESCENT OF MAN

T HE YANOMAMO are a tribal people who dwell in remote forests on
the border of Brazil and Venezuela. Until recent decades, they lived in
a traditional manner, their practices unchanged by missionaries or
other intruders from the civilized world. They dwell in settled villages and
practice agriculture, deriving their staple food from their gardens of plantains,
a kind of large cooking banana. The forest supplies many other prized foods,
such as armadillos, and the delicious grubs, about the size of a mouse, that the
Yanomamo harvest from the pith of palm trees and take home to roast.

The labor required to obtain food is a mere three hours a day. During their ample leisure time, the men snort hallucinogenic drugs prepared from a variety of forest trees while their shamans enter trances from which they communicate with the spirits and recite the myths of the Yanomamo world.

If life is so easy, why then do Yanomamo villages engage in almost continuous warfare with their neighbors? Villages entice others into alliances, bolstered with trade and ritual feasting, for the purpose of defending against or attacking rival coalitions. Not so rarely, the feasts are set-ups for a deadly massacre of the invited guests. The constant warfare carries a serious price. About 30% of all deaths among adult males are due to violence, according to Napoleon Chagnon, the anthropologist who has studied the Yanomamo over many decades.[180] Chagnon found that 57% of people over the age of 40 had lost two or more close relatives—a parent, sibling or child—to a violent death.

The Yanomamo way of life is entirely different from the daily experience of most people in developed economies. Yet it embodies all the institutions that are distinctive of human sociality, including warfare, trade, religion and a defined division of roles between the sexes. Where did these social institutions come from? Do they have biological roots or are they purely cultural? What is it that knits human societies together in the first place?

A possible answer to all these questions, though one for which there is at present no direct evidence, is that human social behavior is rooted in various ways in a genetic template that people have inherited from their primate forebears and that has been adapted throughout evolution to prevailing circumstances.

Those adaptations would seem to include a vigorous expansion of the chimpanzee propensity for territorial defense and aggression against fellow members of the same species. But they must also include a special array of quite different behaviors, ones that enable people to work effectively with others in large, complex societies. In chimpanzee groups, most of the males are related to each other; their common genetic interest is the glue that holds the group together. Humans have developed behaviors that enable even strangers to be treated as kin, a compact basic to all city life. These softer behaviors, which are as much a part of human nature as the propensity to kill and punish, provide the cohesion at the root of civilized societies.

The Dynamics of Primate Societies

The genes that influence human social behavior are inscribed somewhere in the genome but have not yet been recognized. Until they are, the best available guide to them has emerged from the new understanding of how chimpanzee and bonobo societies work. The two ape societies are quite different in character. That of chimps is male dominated and aggressive, whereas bonobos are female dominated and highly conciliatory. Presumably the elements of both kinds of behavior must have existed in the joint human-chimp ancestor from which chimps and bonobos are descended. The social behaviors of the two apes therefore provide invaluable insights into the set of social behaviors that humans too may have inherited from the common primate ancestor.

Chimp society evolved, as might be expected, for the purpose of maximizing the reproductive success of its members. The society's structure seems to be carefully attuned to chimps' general environment, just as the very different social structure of bonobos is appropriate to their environment. Human societies too have a range of different structures, each of which can be seen as a solution to a particular environmental problem. The egalitarian structure of hunter-gatherer societies is well suited to managing the risk of uneven hunting success. The hierarchical structure of settled societies may be a more efficient way of administering surpluses and trade.

The templates for chimp and human social behavior are very similar in a central feature, that of territorial defense and the willingness to solve the problem of a hostile neighboring society by seeking its extermination. But they differ in other critical aspects. Humans have evolved a different relationship between the sexes, based on family units instead of separate male and female hierarchies. These family units require a considerably higher level of trust among males, enabling them to band together for social purposes like warfare with a reasonable degree of confidence that others will not steal their wives. Second, all human societies support institutions not found in the chimp repertoire. These include property rights, a propensity for ceremony, ritual and religion, and elaborate systems of trade and exchange, based on a universal expectation of reciprocity.

Chimpanzee groups, like primitive human societies, are held together by bonds of kin relationships, the evolutionary basis of which is well understood. But kin-bonded societies cannot grow beyond a certain size. Humans, with their special gift of language, have developed ways to knit together large groups of unrelated individuals. One of these binding forces is religion, which may have emerged almost as early as language.

Because of the richness of human culture, it is hard to define the genetic underpinnings of human social behavior. It is much easier to see a set of social behaviors, presumably genetically defined, among our primate cousins. Chimpanzees have been studied in the wild for some 45 years by two pioneers, Jane Goodall at Gombe and Toshisada Nishida at Mahale, both in Tanzania, and by their successors at these and several other sites in Africa. Only in recent years, as the fruit of much arduous research, has the big picture come together. Biologists can now explain many deep features of chimp society and how its components work. The dynamics of chimp society bear directly on the better-concealed game plan of human societies.

Though Goodall at first believed the chimpanzees at Gombe lived in one big happy group, it later became clear, through the stimulus of Nishida's research, that the opposite is the case. Chimps are divided into communities of up to 120 members, which occupy and aggressively defend specific territories.

A chimp community never assembles as a whole. Chimps move around in bands of 20 or so, with shifting membership, in what chimp watchers call a fission-fusion society. The females often feed alone with their offspring or in small nursery parties. A striking parallel with human societies is that these communities are patrilocal, meaning the males stay in their home territory and females move to find mates in neighboring territories. Female chimps generally leave their home communities at the age of puberty and join other communities, whose males tend to find them more attractive than their own.

Most hunter-gatherer societies are patrilocal, in the sense that the wife goes to live with the husband's family. The biological reason is to avoid inbreeding, a problem faced by all social animals. But the almost universal solution in the primate world is matrilocality: the females stay put and the males disperse at puberty. Patrilocality is the exception to the rule, and has probably evolved in only four other primate species besides humans and chimpanzees.[181]

A second unusual feature of a chimp community, but one that chimps also share with people, is a propensity to conduct murderous raids on neighbors. Male chimps not only defend their territory but conduct regular, often lethal, attacks on neighboring communities. This discovery came as a considerable surprise to many biologists and sociologists who had assumed that warfare was a uniquely human phenomenon.

Why do chimps hold and defend territories in the first place? Why do they kill each other? Chimp researchers believe they have been able to unlock the basic logic of chimp social structure, at least in general outline. Chimp society turns out to be matched to the nature of the food supply, which is principally fruit. The trees come into fruit sporadically. They tend to be scattered and do not supply enough fruit for large parties of chimpanzees. Female chimps, needing to sustain themselves and their young, find it more efficient to feed by themselves. They live in home areas, a few square kilometers in size, which they rarely leave. The size of these areas is very important. Females have shorter intervals between births—in other words bear children faster—when their territory is larger, according to an analysis of Gombe chimps by Jennifer Williams and Anne Pusey.

Considering strategies for the male chimps, each could try to achieve reproductive success by guarding one female. But it seems to be more efficient for the males to band together and defend territory that includes a larger number of females. One reason that this makes sense is that the males tend to be related to each other, because of the patrilocal system, and therefore in defending a group of females they are assisting their male relatives' reproductive efforts as well as their own. An individual's kin carry many of the same genes as he does. As the evolutionary biologist William Hamilton pointed out in his doctrine of inclusive fitness, for a person to help get an equivalent number of his kin's genes into the next generation is about as good as propagating his own. This is why genes favoring altruistic behavior have evolved in kin-based societies. The same logic underlies the cohesiveness of ant and bee societies, whose workers, by a special quirk of insect genetics, are more closely related to their sisters than to any daughters they might have. Because of this relationship, the workers have forsaken their own chance to raise children entirely and are content to live as sterile nurses for the queen's children.

In chimpanzee society, males and females do not generally spend much time together except for the purpose of mating. The members of each sex are organized in separate hierarchies. Every adult male demands deference from every female, resorting to immediate violence if a submissive response is not forthcoming. Differences notwithstanding, chimp and human societies serve the same purpose, that of providing males and females appropriate ways of securing their individual reproductive advantage.

At the head of the male hierarchy is the alpha male, who maintains his position by physical strength and, just as importantly, by building alliances with other males. "A dominant male is constantly at risk from opportunistic coalitions formed by lower-ranking individuals and must continually assert his dominance through agonistic display," write John Mitani and colleagues.[182] These tests of leadership, which primatologists sometimes refer to ironically as elections, can occur at any time. Losing an election in chimp society is not a good idea. The loser's defeat may take the form of having personal parts torn off of him and being left for dead. Long rule does not guarantee a peaceful retirement. Ntologi was alpha male at Mahale for 16 years before he was overthrown by a rival coalition and killed.

What is the upside of being alpha male if life is a daily gamble on retaining power, with violent death the only retirement plan on offer? Whether or not chimps ponder this question, evolution has provided the answer: high position in a chimp male hierarchy guarantees that a male will have more matings and more progeny.

This outcome was at first far from obvious to researchers. When females enter their fertile period, they advertise the fact with melon-sized pink swellings on their rear end. They become very gregarious and do their best to mate with every male in the community, with an average of 6 to 8 couplings a day. One female observed by Goodall achieved 50 copulations in one day.[183] The females' purpose, biologists believe, is to confuse paternity. If a male chimp believes there is a chance a baby is his, he is less likely to kill it.

Given this seemingly chaotic mating system, how do high ranking males in fact reap their due rewards of office? First, they do secure more matings, even though rarely exclusive ones. Second, there is the phenomenon of sperm competition. Because of the chimps' multiple mating system, advantage will accrue to the male who can deliver the most sperm and flood out the competition. Hence evolution has favored male chimps with very large testes for their body size. But whether or not the senior males reaped the rewards of rank was unclear until the advent of DNA paternity testing. Julie Constable and colleagues recently reported the results from a 20-year study of chimps at the Kasakela community in Gombe. They found that despite appearances, the system works. The reigning alpha male accounted for 36% of all conceptions, and for 45% if one excludes his close female relatives, with whom conceptions would be avoided.[184] Another 50% of matings were scored by high ranking males. Usually at Gombe there is the alpha male and then two or more other males who count as high ranking.

Most of these conceptions studied by Constable occurred during general free-for-all sexual romps, or "opportunistic matings," as the primatologists call them, suggesting that the alpha males owe of a lot of their fatherhoods to victory in the sperm competition wars.

Like males, female chimps have a hierarchy. It is less discernible, because females spend much of their time feeding alone in their core areas and are not in constant interaction as the males are, but it bears significantly on the females' reproductive success.[185] Low ranking females lose more of their babies than do socially ascendant females. This is partly because socially superior females will sometimes kill the infants of lowly females. The high ranking Passion and her daughter Pom snatched and ate the babies of their neighbors at Gombe, perhaps to discourage trespass on their feeding areas.

What makes one female dominant over another is not yet clear, but in general terms rank in chimpanzee society seems to depend a lot on one's mother's status. Flo, a high ranking and sexually attractive female, was the mother of Figan, who was alpha male in his Gombe community for 10 years (his reign date was 1971–1981), as well as of Fifi, who became dominant female. Fifi helped her firstborn son Freud take his first steps to power by intervening on his side when, as an adolescent, he started to establish dominance over the females. Freud was alpha male from 1994 to 1998, when he fell sick with mange and was deposed by his younger brother Frodo.

Historians attribute dynastic wars among people to all kinds of complex motives, from glory to territorial gain to spread of religion. Chimpanzees' intentions, unobscured by such rationalizations, can be judged strictly by their results. It's all about reproductive advantage. Each player acts so as to get as many descendants as possible into the next generation. The males try to ascend the male dominance hierarchy so as to mate with as many females as possible. The females seek out the best feeding areas so as to bear as many surviving children as possible. The ultimate objective is simple, but in a complex society each individual must act in many intricate ways to achieve it.

Presumably chimps' social behavior is genetically shaped, but like human societies they have culture too, in the sense of learned behavior that varies from one chimp community to another. In a recent survey of seven long term chimp studies, Andrew Whiten and colleagues identified 39 behaviors that differed from one community to another without obvious ecological explanation.[186] All chimpanzee communities habitually use tools, but the use of tools varies widely from one chimp community to another. Chimps in the Tai forest in the Ivory Coast use stones as hammers to open nuts; Gombe's chimps have never learned or invented this useful art. Not a single case of habitual tool use has yet been observed among bonobos.[187] That suggests that chimps have a genetic propensity to use tools and bonobos do not.

If the variations between chimp communities are mostly due to culture, the constant features of chimp social behavior are probably framed by genes. And presumably a shift in that genetic framework for social behavior explains the difference between chimpanzees' social arrangements and those of their cousins, the bonobos, from whom they have been separated for some two million years.

The Bonobo Alternative

Bonobos are so similar to chimpanzees in physical appearance that it took biologists many years to recognize that they are a separate species. Their behavior, however, is very different. Unlike in chimp societies, where males may violently coerce females to respect them, in bonobo land the females run the show. They manage this feat by forming close alliances with each other and facing down any male who tries to interfere in their affairs. Because of their dominance, they have managed to banish infanticide, the worst fear of female chimpanzees.

Bonobos have captured the attention of their human observers because they use sex not just for reproduction but also as a social greeting and general reconciliation technique. Bonobo sexual physiology has a small but socially critical difference from that of chimpanzees. Male chimpanzees seem to be able to tell, probably by smell, the almost exact time when a female is ovulating, setting off fierce competition for her favors. But bonobo ovulation, as with humans, is concealed. The males, who get to have sex with the females almost all the time anyway, do not enter into ferocious competition with each other because the goalpost, as it were, is no longer in sight. Bonobo social arrangements do a superb job, from the female point of view, of making paternity utterly obscure.

Bonobo communities are considerably less aggressive to each other than are those of chimpanzees. There are no border patrols by groups of males looking for trouble. Groups from two communities have even been observed mingling peacefully, to the astonishment of chimpanzee biologists.

Why is bonobo behavior so different from that of chimpanzees? The answer seems to be that bonobo society has evolved in adaptation to a subtle but profound difference in the bonobo environment. Following is the analysis offered by Richard Wrangham, a chimpanzee expert at Harvard University, based in part on the observations of bonobos by the Japanese primatologist Takayoshi Kano and his colleagues.

At first sight, there is no obvious ecological difference between chimp habitat and bonobo habitat. They both live in tropical rain forests, although the chimps inhabit some more open woodland as well. Chimps are found all across tropical Africa, from the west coast to the east, but bonobos live south of the Zaire river, and chimps live north of it.

The river is a barrier, and south of the river there are no gorillas. Gorillas are voracious eaters of herbaceous plants. The chimps north of the river, who share their territory with gorillas, eat only fruit, leaving the herbs for the gorillas. But the bonobos south of the river eat both, and their teeth are specially adapted for shearing herbs.

This difference in diet has far reaching consequences. Female chimpanzees forage alone, in their core feeding areas, because that is the most efficient way to get enough to eat. But since bonobo forests have more sources of food, bonobos can travel in larger parties with a more stable membership. This gives the females the opportunity to bond together, which they do with the usual bonobo social lubricant—plenty of sex. "Party stability, in other words, produced female power," Wrangham says.[188]

With both chimps and bonobos, social structure is designed so as let each species make best use of its environment. Considerable genetic change must have occurred for bonobos to evolve from a chimpanzee-like ances-tor.[189] Bonobo males had to become less aggressive, females more adept at forming coalitions powerful enough for their hierarchy to control that of the males.

Although the point cannot yet be proved, it seems more likely that bono-bos are descended from chimpanzees, rather than the other way around. Still, both are descended from the joint ancestor of chimps and humans, and the joint ancestor presumably included both chimplike and bonobolike features in its behavioral template. That makes it easier to understand how humans came by their contradictory impulses of aggression and concilia-tion. "Being both systematically more brutal than chimps and more em-pathic than bonobos, we are by far the most bipolar ape," writes the primatologist Frans de Waal. "Our societies are never completely peaceful, never completely competitive, never ruled by sheer selfishness, and never perfectly moral."[190]

The Costs and Benefits of Warfare

Besides being well adapted or designed for their environments, chimp and human societies possess another salient feature in common, that of a strong propensity to kill their own kind. A willingness to kill members of one's own species is apparently correlated with high intelligence. It may be that chimps and people are the only species able to figure out that the extra effort re-quired to exterminate an opponent will bring about a more permanent solu-tion than letting him live to fight another day.

Military skills are probably underappreciated as a biological phenome-non, but in their own way are just as remarkable a human adaptation as is the artistic ability of the Upper Paleolithic cave painters. Warfare of the hu-man kind has many levels of complexity and at its highest is an integral com-ponent of statecraft. At the lower end of the scale, however, it overlaps closely in both tactics and goals with the chimpanzee variety.

Chimp warfare takes the form of bands of males who patrol the borders of their territory, looking for an individual of the neighboring community who has been rash enough to feed alone. Occasionally they make raids deep into enemy territory. "Behavior during patrols is striking and unusual," writes the primatologist John Mitani. "Males are silent, tense and wary. They move in tight file, often pause to look and listen, sometimes sniff the ground, and show great interest in chimpanzee nests, dung, and feeding remains." Just like human raiders, they are tense and nervous.[191]

Chimpanzees carefully calculate the odds and seek to minimize risk, a very necessary procedure if one fights on a regular basis. They prefer to attack an isolated individual and then retreat to their own territory. If they encounter an opposing patrol they will assess the size of their opponents' party and retreat if outnumbered. Researchers have confirmed this behavior by playing the call of a single male through a loudspeaker to chimp parties of various sizes. They find that the chimps will approach as long as they number three or more; parties of two will slink away. Three against one is the preferred odds: two to hold the victim down and a third to batter him to death.

The raid is also the principal kind of warfare conducted by primitive human societies. Yanomamo raids too are carefully calculated to minimize risk. "The objective of the raid is to kill one or more of the enemy and flee without being discovered," writes Napoleon Chagnon.[192]

Warfare is a bond that separates humans and chimps from all other species. "Very few animals live in patrilineal, male-bonded communities wherein females routinely reduce the risks of inbreeding by moving to neighboring groups to mate," write Richard Wrangham and Dale Peterson. "And only two animal species are known to do so with a system of intense, male-initiated territorial aggression, including lethal raiding into neighboring communities in search of vulnerable enemies to attack and kill. Out of four thousand mammals and ten million or more other animal species, this suite of behaviors is known only among chimpanzees and humans."

In their resort to warfare, both chimps and human societies, at least those like the Yanomamo, have the same essential motivation. The chimps are defending fruit tree territory for the females, for their own reproductive advantage. The Yanomamo have the same idea in mind. Capture of women is seldom the prime reason for a raid but is an expected side benefit. A captured woman is raped by all members of the raiding party, then by everyone back home who wishes to do so, and is then given to one of the men as a wife.

But the real reproductive advantage of participating in a raid derives from the prestige of killing an enemy. When a man has killed someone he must perform a ritual purification called a *unokaimou* to avert retaliation by the soul of his victim. Those who have undergone this ritual are called *unokai*, and it is well known who they are. The *unokais*, Chagnon found, have on average 2.5 times as many wives as men who have not killed, and over three times as many children.

Chagnon's study of the Yanomamo is unusual because he has studied them over such a long period of time. Despite the thoroughness of his fieldwork, some anthropologists have been reluctant to accept his conclusions, resisting the idea that violence could be reproductively rewarding. One critic, Marvin Harris, suggested that Yanomamo warfare was driven by a scarcity of protein. Chagnon describes the Yanomamo's reception of this idea. "I explained Harris's theory of their warfare to the Yanomamo: 'He says you are fighting over game animals and meat, and insists that you are not fighting over women.' They laughed at first, and then dismissed Harris's view in the following way: '*Yahi yamako buhii makuwi, suwa kaba yamako buhii barowo!*' ('Even though we do like meat, we like women a whole lot more!')"[193]

Why would the Yanomamo pursue a way of life with such a high risk of violent death? The greater reproductive advantage of being a *unokai* is the obvious answer, a motivation that of course need not be conscious. Chimpanzees provide the same answer as the *unokais*, and bear an almost identical cost. In Gombe, some 30% of adult males died from aggression, the same toll as among the Yanomamo. A man or chimp may die defending his territory, but he still has a chance of propagating his genes. The males who may profit from his sacrifice are his relatives and carry many of the same genes. Raiders will be rewarded and have sons of similar character. That is the logic of patrilocality.

The Efficacy of Primitive Warfare

A propensity for warfare is prominent among the suite of behaviors that people and chimpanzees have inherited from their joint ancestor. The savagery of wars between modern states has produced unparalleled carnage. Yet the common impression that primitive peoples, by comparison, were peaceful and their occasional fighting of no serious consequence is incorrect.

Warfare between pre-state societies was incessant, merciless, and conducted with the general purpose, often achieved, of annihilating the opponent. As far as human nature is concerned, people of early societies seem to have been considerably more warlike than are people today. In fact, over the course of the last 50,000 years, the human propensity for warfare has probably been considerably attenuated.

"Peaceful pre-state societies were very rare; warfare between them was very frequent, and most adult men in such groups saw combat repeatedly in a lifetime," writes Lawrence H. Keeley, an archaeologist at the University of Illinois at Chicago. Primitive warfare was conducted not by arrays of troops on a formal battlefield, in the western style, but by raids, ambushes and surprise attacks. The numbers killed in each raid might be small, but because warfare was incessant, the casualties far exceeded the losses of state societies when measured as a percentage of population. "In fact, primitive warfare was much more deadly than that conducted between civilized states because of the greater frequency of combat and the more merciless way it was conducted. Primitive war was very efficient at inflicting damage through the destruction of property, especially means of production and shelter, and inducing terror by frequently visiting sudden death and mutilating its victims."[194]

Keeley's conclusions are drawn from the archaeological evidence of the past, including the Upper Paleolithic period, and from anthropological studies of primitive peoples. These include three groups of foragers that survived until recent times—the !Kung San, Eskimos and Australian aborigines—as well as tribal farmers such the Yanomamo of Brazil and the pig and yam cultivating societies of New Guinea.

To minimize risk, primitive societies chose tactics like the ambush and the dawn raid. Even so, their casualty rates were enormous, not least because they did not take prisoners. That policy was compatible with their usual strategic goal: to exterminate the opponent's society. Captured warriors were killed on the spot, except in the case of the Iroquois, who took captives home to torture them before death, and certain tribes in Colombia, who liked to fatten prisoners before eating them.

Warfare was a routine occupation of primitive societies. Some 65% were at war continuously, according to Keeley's estimate, and 87% fought more than once a year.[195] A typical tribal society lost about 0.5% of its population in combat each year, Keeley found. Had the same casualty rate been suffered by the population of the twentieth century, its war deaths would have totaled two billion people.

On the infrequent occasions when primitive societies fought pitched battles, casualty rates of 30% or so seem to have been the rule. A Mojave Indian war party was expected to lose 30% of its warriors in an average battle. In a battle in New Guinea, the Mae Enga tribe took a 40% loss. At Gettysburg, by comparison, the Union side lost 21%, the Confederates 30%.

An archaeologist, Steven LeBlanc of Harvard University, recently reached similar conclusions to Keeley after an independent study. "We need to recognize and accept the idea of nonpeaceful past for the entire time of human existence," he writes. "Though there were certainly times and places during which peace prevailed, overall, such interludes seem to have been short-lived and infrequent. . . . To understand much of today's war, we must see it as a common and almost universal human behavior that has been with us as we went from ape to human."[196]

Primitive warriors were highly proficient soldiers, Keeley notes. When they met the troops of civilized societies in open battle, they regularly defeated them despite the vast disparity in weaponry. In the Indian wars, the U.S. Army "usually suffered severe defeats" when caught in the open, such as by the Seminoles in 1834, and at the battle of Little Bighorn. In 1879 the British army in South Africa, equipped with artillery and Gatling guns, was convincingly defeated by Zulus armed mostly with spears and ox-hide shields at the battles of Isandlwana, Myer's Drift and Hlobane. The French were seen off by the Tuareg of the Sahara in the 1890s. The state armies prevailed in the end only through larger manpower and attritional campaigns, not by any superior fighting skill.

How did the warriors of primitive societies get to be so extraordinarily good at their craft? By constant practice during some 50,000 years of unrestrained campaigning. Even in the harshest possible environments, where it was struggle enough just to keep alive, primitive societies still pursued the more overriding goal of killing one another. The anthropologist Ernest Burch made a careful study of warfare among the Eskimos of northwest Alaska. He learned, LeBlanc reports, "that coastal and inland villages were often located with defense in mind—on a spit of land, or adjacent to thick willows, which provided a barrier to attackers. Tunnels were sometimes dug between houses so people could escape surprise raids. Dogs played an important role as sentinels. The goal in all warfare among these Eskimos was annihilation, Burch reported, and women and children were normally not spared, nor were prisoners taken, except to be killed later. Burning logs and bark were thrown into houses to set them on fire and to force the inhabitants out, where they could be killed. Burch's study reveals that the surprise dawn raid was the typical and preferred war tactic, but open battles did occur."

Both Keeley and LeBlanc believe that for a variety of reasons anthropologists and their fellow archaeologists have seriously underreported the prevalence of warfare among primitive societies. "While my purpose here is not to rail against my colleagues, it is impossible to ignore the fact that academia has missed what I consider to be some of the essence of human history," writes LeBlanc. "I realized that archaeologists of the postwar period had artificially 'pacified the past' and shared a pervasive bias against the possibility of prehistoric warfare," says Keeley.

Keeley suggests that warfare and conquest fell out of favor as subjects of academic study after Europeans' experiences of the Nazis, who treated them, also in the name of might makes right, as badly as they were accustomed to treating their colonial subjects. Be that as it may, there does seem a certain reluctance among archaeologists to recognize the full extent of ancient warfare. Keeley reports that his grant application to study a nine-foot-deep Neolithic ditch and palisade was rejected until he changed his description of the structure from "fortification" to "enclosure." Most archaeologists, says LeBlanc, ignored the fortifications around Mayan cities and viewed the Mayan elite as peaceful priests. But over the last 20 years Mayan records have been deciphered. Contrary to archaeologists' wishful thinking, they show the allegedly peaceful elite was heavily into war, conquest and the sanguinary sacrifice of beaten opponents.

Archaeologists have described caches of large round stones as being designed for use in boiling water, ignoring the commonsense possibility that they were slingshots. When spears, swords, shields, parts of a chariot and a male corpse dressed in armor emerged from a burial, archaeologists asserted that these were status symbols and not, heaven forbid, weapons for actual military use. The large number of copper and bronze axes found in Late Neolithic and Bronze Age burials were held to be not battle axes but a form of money. The spectacularly intact 5,000-year-old man discovered in a melting glacier in 1991, named Ötzi by researchers, carried just such a copper axe. He was found, Keeley writes dryly, "with one of these moneys mischievously hafted as an ax. He also had with him a dagger, a bow, and some arrows; presumably these were his small change."

Despite the fact that the deceased was armed to the teeth, archaeologists and anthropologists speculated that he was a shepherd who had fallen asleep and frozen peacefully to death in a sudden snowstorm, or maybe that he was a trader crossing the Alps on business. Such ideas were laid to rest when an X-ray eventually revealed an arrowhead in the armed man's chest. "In spite of a growing willingness among many anthropologists in recent years to accept the idea that the past was not peaceful," LeBlanc comments, "a lingering desire to sanitize and ignore warfare still exists within the field, Naturally the public absorbs this scholarly bias, and the myth of a peaceful past continues."

If primitive societies of the historic past were heavily engaged in warfare, it seems quite possible that their distant ancestors were even more aggressive. A genetic discovery made as part of a study of mad cow disease lends some credence to this idea.

The Skeleton in the Human Past

Among the least appetizing aspects of primitive warfare is cannibalism. Cannibalism implies the existence of warfare since the victims do not voluntarily place themselves on the menu. Anthropologists and archaeologists have long resisted the idea that cannibalism took place in the peaceful past. In his 1979 book *Man-Eating Myth*, William Arens, an anthropologist at the State University of New York at Stony Brook, argued that there was no well attested case of cannibalism and that most reports of it were propaganda made by one society to establish its moral superiority to another. Christy G. Turner, an archaeologist at Arizona State University, met only disbelief when he first proposed that the cut, burned, and defleshed bones of 30 individuals at a site occupied by Anasazi Indians were the remains of an ancient cannibal feast. His critics attributed the cuts on such bones to scavenging animals, funerary practices, the roof falling in—anything but anthropophagy.

Though some accounts of cannibalism may well have been fictive, Turner and Tim White of the University of California at Berkeley have now found cannibalized human remains at 25 sites in the American southwest. Turner believes these are the work of Anasazi Indians who dominated the area between AD 900 and 1700 and used cannibalism as an instrument of social control. Cannibalism has been reported from Central and South America, Fiji, New Zealand and Africa. The Aztecs made a state practice of sacrificing captives and their civilization has furnished a recipe for human stew. A common belief that accompanies ritual cannibalism is the notion that by eating particular parts of the victim, often a slain warrior, the consumer absorbs his strength or courage. The frequency of reports of cannibalism by societies in all regions of the world suggests, Keeley concludes, "that, while hardly the norm, ritual consumption of some part of enemy corpses was by no means rare in prestate warfare."

Could cannibalism in fact have been so widespread and so deeply embedded in human practice as to have left its signature in the human genome? This gruesome possibility has emerged from the work of English researchers trying to assess the likely extent of the outbreak of mad cow disease among Britons who had eaten tainted beef. Mad cow disease belongs to a group of brain-eroding pathologies caused by misshapen brain proteins known as prions. Contrary to the expectations of British agricultural officials, prions can cross species barriers; cow prions, which rot cows' brains, can also rot human brains if the cow's neural tissue is eaten.

Even more effective at rotting the human brain are human prions. People are at risk of exposure to human prions when they eat other people's brains. This was a regular practice among the Fore of New Guinea who, sometime around the year 1900, adopted the novel funerary practice of having women and children eat the brains of the dead. By about 1920, the first case of a brain-wasting disease they called *kuru* appeared.

A very similar disease, called Creutzfeldt-Jakob disease or CJD, occurs at low incidence in many populations of the world. CJD is caused after a spontaneous mutation causes brain cells to make the misshapen form of the protein instead of the normal form. *Kuru* presumably started when the brain of a deceased Fore with a natural case of CJD was eaten by his relatives.

Once *kuru* got a foothold in the Fore population, the disease progressed relentlessly until some villages became almost devoid of young adult women.

The epidemic quickly subsided after Australian administrative authorities banned the Fore's mortuary feasts in the 1950s.

A research team led by Simon Mead of University College, London, recently looked at the genetics of Fore women aged over 50. All these survivors had attended many funeral feasts and presumably must have possessed some genetic protection against the disease. Mead's team analyzed the DNA of their prion protein gene and found that more than 75% had a distinctive genetic signature.[197] Every person in Britain infected with mad cow disease, on the other hand, had the opposite genetic signature.*

Having identified this protective signature, Mead's team then analyzed other populations around the world. They found that every ethnic group they looked at possessed the signature with the exception of the Japanese, who had a protective signature of their own at a different site in the gene.

Various genetic tests showed that the protective signature was too common to have arisen by chance, and must have been amplified through natural selection. Other tests suggested the signature was very ancient and was probably present in the human population before it dispersed from Africa. Under this scenario the Japanese presumably lost the signature through the process known as genetic drift, but developed a new one instead because it was so necessary.

So why has the British epidemic of mad cow disease proved not nearly so deadly in that nation of beef-eaters as was initially feared? It seems that Britons have been in part protected by their ancient cannibal heritage. That the British and other world populations have maintained the protective signature many generations after their last cannibal feast is an indication of how widespread cannibalism may have been in the ancestral human population and its worldwide descendants. The frequency of cannibalism in turn attests to the prevalence of warfare among the earliest human populations.

"There is an innate predisposition to manufacture the cultural apparatus of aggression, in a way that separates the conscious mind from the raw biological processes that the genes encode," writes the biologist Edward O.

---

*At a particular region of the prion protein's gene, known as codon 129, the gene exists in two forms. Since a person has two copies of each gene, one from each parent, it's possible to inherit 1) each of the two forms, one from each parent, or 2) two copies of the same form. Having two copies of the same form of the prion gene turns out to be a risk factor for mad cow and related diseases; having two different forms is protective.

Wilson. "Culture gives a particular form to the aggression and sanctifies the uniformity of its practice by all members of the tribe."[198] The genes supply the motivation for warfare, Wilson is saying, in humans as they do in chimps, but people, blessed with the power of language, look for some objective cause of war. A society psychs itself up to go to war by agreeing that their neighbors have wronged them, whether by seizing property or failing to deliver on some promise. Religious leaders confirm that the local deity favors their cause and off go the troops.

The human predisposition for socially approved aggression falls into a quite different category from that of individual aggression. Bellicose individuals usually get themselves locked up in jail for long periods or, in primitive societies, social sanction is given to having them killed. Individual aggression is seldom a good strategy for propagating one's genes. But socially approved aggression—that is, warfare—can be. A predisposition to warfare does not mean war is inevitable since the predisposition is only executed in certain contexts. The warlike Vikings of the tenth century became the peaceful Scandinavians of the twentieth.

Among forager societies, warfare can benefit the victor, by expanding territory and increasing reproductive success. That is the conclusion that archaeologists and anthropologists have been so anxious to avoid endorsing, because it seems to offer a justification for war, even a glorification of it. But by playing down the prevalence of warfare in the past they have obscured the important and surprising fact adduced by Keeley, that modern societies have succeeded in greatly reducing the frequency of warfare.

On the assumption that warfare was an incessant preoccupation of early human existence, the picture of the Upper Paleolithic era that specialists have so far constructed seems strangely incomplete. What does it mean to say that the Aurignacian culture was succeeded by the Gravettian? That the makers of the Aurignacian tool kit woke up one morning and decided thenceforward they would all do things the Gravettian way? Or that after many sanguinary battles people bearing the Gravettian culture ousted those following the Aurignacian? When the Last Glacial Maximum made northern latitudes uninhabitable and the glaciers pushed their populations south, is it likely they were welcomed with open arms by the southerners whose territory they invaded? If warfare was the normal state of affairs, it would have shaped almost every aspect of early human societies.

Warfare is a dramatic and distinctive feature of history, and it thoroughly overshadows an even more remarkable feature of human societies. This feature, the polar opposite of war, is the unique human ability to cooperate with others, and specifically with unrelated individuals. Social organisms like bees and ants form groups centered around members who are related to each other and have a common genetic interest. So do people to some extent when organized in tribal societies. But humans have extended sociality far beyond the extended family or tribe and have developed ways for many unrelated individuals to cooperate in large, complex, cohesive societies.

The uniquely human blend of sociality was not easily attained. Its various elements evolved over many years. The most fundamental, a major shift from the ape brand of sociality, was the human nuclear family, which gave all males a chance at procreation along with incentives to cooperate with others in foraging and defense. A second element, developed from an instinct shared with other primates, was a sense of fairness and reciprocity, extended in human societies to a propensity for exchange and trade with other groups. A third element was language. And the fourth, a defense against the snares of language, was religion. All these behaviors are built on the basic calculus of social animals, that cooperation holds more advantages than competition.

The Evolutionary Basis of Social Behavior

Though we take the necessities of social behavior for granted, group living in the animal world is highly unusual. In fact even the most rudimentary forms of sociality have long been a puzzle for biologists to explain in terms of evolutionary theory.

The reason is that a society serves no purpose unless members help one another, yet any effort an individual makes assisting others takes away from investment in his own offspring and reproductive success. If altruists have fewer children, altruistic behavior will be eliminated by natural selection. Yet without altruism there is no benefit to living in a society. How therefore can social behavior ever have evolved?

Evolutionary biologists have developed a reasonably good account of how social behavior may have emerged in groups of closely related individuals, in a theory about what is known as inclusive fitness. Another theory, that of reciprocal altruism, explains how behavior could have evolved for helping even unrelated people, or at least those who can be expected to reciprocate the favor at a later time.

Why will a bee sacrifice its life in the hive's defense? Why should a worker ant embrace sterility and devote her life to raising the queen's offspring? The late William Hamilton made a major addition to Darwin's theory in showing how altruism, at least toward one's own kin, makes evolutionary sense. Darwinian fitness, defined as reproductive success, is all about getting as many of one's own genes as possible into the next generation. Hamilton's insight was that the notion of Darwinian fitness should properly be expanded to include the genes one shares with one's kin. Since these shared genes are the same, being inherited from the same parent, grandparent or great-grandparent, then helping get those into the next generation is as good as transmitting one's own.

This notion of expanded fitness, or inclusive fitness as Hamilton called it, predicts that individuals will have a special interest in promoting the survival of children, full siblings and parents, with all of whom they have about 50% of their genes in common, and a substantial though lesser interest in the survival of grandchildren, nephews and nieces, half siblings, grandparents, and aunts and uncles, with whom they share 25% of their genes.

To maximize their inclusive fitness, individuals must restrain their own competitive behavior and make some degree of self-sacrifice on behalf of kin—in other words, develop social behavior. Thus altruists can be inclusively fitter than non-altruists and their genes, under certain conditions, will spread. Hamilton's theory of inclusive fitness explains many otherwise puzzling features of social organisms, such as the self-sacrificing behavior of social insects like bees, ants and termites. It also helps explain why chimp communities and human tribal societies are organized along kinship lines.

It can take extreme circumstances to make evident the survival value of human kinship ties. Some 51% of the 103 *Mayflower* pioneers in the Plymouth colony perished after their first winter in the New World. It turns out that the survivors had significantly more relatives among other members of the colony than did those who died. Among the Donner party, a group of 87 people stranded in the Sierra Nevada in the winter of 1846, only 3 of 15 single young men survived, whereas men who survived had an average of 8.4 family members with them.[199]

But kinship alone seems to have limited power as a cohesive social force. Napoleon Chagnon, in his study of the Yanomamo, noticed that as village populations grew larger, the average degree of relatedness would decrease. The population would then split, usually along kinship lines, with the result that people within the two smaller groups would be more highly related to each other. "Kinship-organized groups can only get so large before they begin falling apart," Chagnon writes. Disputes break out over the usual things—sexual trysts, infidelity, snide comments or veiled insults. "As villages grow larger, internal order and cooperation become difficult, and eventually factions develop: Certain kin take sides with each other, and social life becomes strained. There appears to be an upper limit to the size of a group that can be cooperatively organized by the principles of kinship, descent and marriage, the 'integrating' mechanisms characteristically at the disposal of primitive peoples."[200]

In most Yanomamo villages, members are on average related to each other more closely than half-cousinship.[201] But nontribal societies are a lot larger, as if some new cohesive factor has to come into play if a community is to outgrow the organizational limits imposed by kinship. Recent human history, Chagnon writes, could be viewed as a struggle to overcome these limits: "Many general discussions of our social past as hunters and early cultivators allude to the 'magic' numbers of 50 to 100 as the general community size within which our recent cultural and biosocial evolution occurred, a maximal community size that was transcended only in the very recent past—within the last several thousand years."[202]

One principle that biologists think may help explain larger societies, both human and otherwise, is that of reciprocal altruism, the practice of helping even a nonrelated member of society because they may return the favor in future. A tit-for-tat behavioral strategy, where you cooperate with a new acquaintance, and thereafter follow his strategy toward you (retaliate if he retaliates, cooperate if he cooperates), turns out to be superior to all others in many circumstances. Such a behavior could therefore evolve, providing that a mechanism to detect and punish freeloaders evolves in parallel; otherwise freeloaders will be more successful and drive the conditional altruists to extinction.

Conditional or tit-for-tat altruism cannot evolve in just any species. It requires members to recognize each other and have long memories, so as to be able to keep tally. A species that provides a shining example of reciprocal altruism is none other than the vampire bat. The bats, found in South America, hang out in colonies of a dozen or so adult females with their children. They feed by biting a small incision in the skin of sleeping animals, nowadays mostly cattle or horses, and injecting a special anticoagulant named, naturally enough, draculin. But their blood collection drives are not always successful. On any given night a third of the young bats and 7% of the adults are unsuccessful, according to a study by Gerald W. Wilkinson of the University of Maryland.

This could pose a serious problem because vampire bats must feed every three days, or they die. The colony's solution, Wilkinson found, is that successful bats regurgitate blood to those who went hungry. Bats are particularly likely to donate blood to their friends, with whom they have grooming relationships, to those in dire need, and to those from whom they received help recently.[203] The vampires' reciprocal altruism must be particularly effective since the bats, despite the risk of death after three bloodless nights, can live for 15 years.

If social altruism has evolved among vampire bats, there is no reason why it could not also emerge among primates. And indeed it can be seen at work in the coalitionary politics of male chimpanzees, where the alpha male depends on allies to preserve his dominance of the male hierarchy. The biologist Robert L. Trivers, who first showed how reciprocal altruism could be favored by natural selection, suggested that in people a wide range of sophisticated behaviors grew up around it, including cheating (failure to return an altruistic favor to the giver), indignation at cheating, and methods to detect cheating.[204]

Many common emotions can be understood as being built around the expectation of reciprocity and the negative reaction when it is made to fail. If we like a person, we are willing to exchange favors with them. We are angry at those who fail to return favors. We seek punishment for those who take advantage of us. We feel guilty if we fail to return a favor, and shame if publicly exposed. If we believe someone is genuinely sorry about a failure to reciprocate, we trust them. But if we detect they are simulating contrition, we mistrust them.[205]

The instinct for reciprocity, and the cheater-detection apparatus that accompanies it, seem to be the basis for a fundamental human practice, that

of trade with neighboring groups. Long distance trade is one of the characteristic behaviors of the human societies that emerged in the Upper Paleolithic age starting some 50,000 years ago. Tribal societies developed trading
systems of considerable sophistication. The Yit Yoront, a foraging society of
northern Australia, lived until recently in the Stone Age. One of their most
necessary possessions, used in everything from hunting to wood-gathering,
were hafted stone axes. But they lived on an alluvial coast and the nearest
stone quarry was four hundred miles away. How did they acquire their pol-
ished stone axes? They made a product much in demand with their neighbors to the south, spears tipped with the barbs of stingrays. The spears were
traded inland, through a long line of trading partners, being exchanged at
each stage for a varying number of stone axes. The spear/axe exchange rate
was sufficient at each trading post to push stone axes northward and pull
barb-tipped spears southward.[206]

Trade is a foundation of economic activity because it gives the parties to
a transaction a strong incentive to specialize in making the items that the
other finds valuable. But trade depends on trust, on the decision to treat a
total stranger as if he were a member of the family. Humans are the only
species to have developed such a degree of social trust that they are willing
to let vital tasks be performed by individuals who are not part of the family.
This set of behaviors, built around reciprocity, fair exchange and the detection of cheaters, has provided the foundation for the most sophisticated urban civilizations, including those of the present day.

Reciprocity, and an ability to calculate the costs and benefits of cooperation, underpin our social life, writes the economist Paul Seabright, "making
it reasonable for us to treat strangers as though they were honorary relatives
or friends." It is remarkable that this behavior evolved at a time when primitive warfare was at its most intense and people had every reason to regard
strangers with deep suspicion. Strangers can still be dangerous, yet in the
right circumstances we habitually trust them. "The knowledge that most
people can be trusted much of the time to play their part in the complex
web of social cooperation has had dramatic effects on the psychology of our
everyday life," Seabright says, making it possible "to step nonchalantly out
of the front door of a suburban house and disappear into a city of ten million strangers."[207] Without this innate willingness to trust strangers, human

societies would still consist of family units a few score strong, and cities and great economies would have had no foundation for existence.

How might this greater level of trust have arisen? Two hormones, known as oxytocin and vasopressin, are emerging as central players in modulating certain social behaviors in the mammalian brain. The hormones are generated in the pituitary gland at the base of the brain and have effects both on the body and in the brain. Oxytocin induces both labor in childbirth and the production of milk. Its effects on the mind, at least in experimental animals, have the general property of promoting affiliative or trusting behavior, lowering the natural resistance that animals have to the close proximity of others.

So what does oxytocin do in people? Researchers at the University of Zurich have found that it substantially increases the level of trust. Oxytocin, they say, "specifically affects an individual's willingness to accept social risks arising through interpersonal interactions." The findings emerged from giving subjects a sniff of oxytocin before playing a game that tested trusting behavior.[208]

If the biological basis of trusting behavior is mediated in this manner, the degree of trust could easily be ratcheted up or down in the course of human evolution by genetic changes that either increased individuals' natural production of the hormone or enhanced the brain's response to it. Thus hunter-gatherers might have a genetically lower response to oxytocin while city-dwellers would have evolved a greater sensitivity. Whatever the exact mechanism, it is easy to see how greater levels of trust might have evolved at various stages in human evolution, given that there is a biological basis for the behavior.

Trust is an essential part of the social glue that binds people together in cooperative associations. But it increases the vulnerability to which all social groups are exposed, that of being taken advantage of by freeloaders. Freeloaders seize the benefits of social living without contributing to the costs. They are immensely threatening to a social group because they diminish the benefits of sociality for others and, if their behavior goes unpunished, they may bring about the society's dissolution.

Human societies long ago devised an antidote to the freeloader problem. This freeloader defense system, a major organizing principle of every society, has assumed so many other duties that its original role has been lost sight of. It is religion.

The Evolution of Religion

The essence of religion is communal: religious rituals are performed by assemblies of people. The word itself, probably derived from the Latin *religare*, meaning to bind, speaks to its role in social cohesion. Religious ceremonies involve emotive communal actions, such as singing or dancing, and this commonality of physical action reinforces the participants' commitment to the shared religious views.

The propensity for religious belief may be innate since it is found in societies around the world. Innate behaviors are shaped by natural selection because they confer some advantage in the struggle for survival. But if religion is innate, what could that advantage have been?

No one can describe with certainty the specific needs of hunter-gatherer societies that religion evolved to satisfy. But a strong possibility is that religion coevolved with language, because language can be used to deceive, and religion is a safeguard against deception. Religion began as a mechanism for a community to exclude those who could not be trusted. Later, it grew into a means of encouraging communal action, a necessary role in hunter-gatherer societies that have no chiefs or central authority. It was then co-opted by the rulers of settled societies as a way of solidifying their authority and justifying their privileged position. Modern states now accomplish by other means many of the early roles performed by religion, which is why religion has become of less relevance in some societies. But because the propensity for religious belief is still wired into the human mind, religion continues to be a potent force in societies that still struggle for cohesion.

A distinctive feature of religion is that it appeals to something deeper than reason: religious truths are accepted not as mere statements of fact but as sacred truths, something that it would be morally wrong to doubt. This emotive quality suggests that religion has deep roots in human nature, and that just as people are born with a propensity to learn the language they hear spoken around them, so too they may be primed to embrace their community's religious beliefs.

Can the origin of religion be dated? A surprising answer is yes, if the following argument is accepted. Like most behaviors that are found in societies throughout the world, religion must have been present in the ancestral human population before the dispersal from Africa 50,000 years ago. Although religious rituals usually involve dance and music, they are also very verbal, since the sacred truths have to be stated. If so, religion, at least in its modern form, cannot pre-date the emergence of language. It has been argued earlier that language attained its modern state shortly before the exodus from Africa. If religion had to await the evolution of modern, articulate language, then it too would have emerged shortly before 50,000 years ago.

If both religion and language evolved at the same time, it is reasonable to assume that each emerged in interaction with the other. It is easy enough to see why religion needed language, as a vehicle for the sharing of religious ideas. But why should language have needed religion?

The answer may have to do with the instinct for reciprocal altruism that is a principal cohesive force in human society, and specifically with its principal vulnerability, the freeloaders who may take advantage of the system without returning favors to others. Unless freeloaders can be curbed, a society may disintegrate, since membership loses its advantages. With the advent of language, freeloaders gained a great weapon, the power to deceive. Religion could have evolved as a means of defense against freeloading. Those who committed themselves in public ritual to the sacred truth were armed against the lie by knowing that they could trust one another.

The anthropologist Roy Rappaport argued that sanctified statements were early societies' antidote to the misuse of the newly emerged powers of language. "This implies that the idea of the sacred is as old as language," he wrote, "and that the evolution of language and of the idea of the sacred were closely related, if not bound together in a single mutual causal process." The emergence of the sacred, he suggested, "possibly helped to maintain the general features of some previously existing social organization in the face of new threats posed by an ever-increasing capacity for lying."[209]

For early societies making the first use of language, there had to be some context in which statements were reliably and indubitably true. That context, in Rappaport's view, was sanctity. This feature has been retained to a considerable degree in modern religions, which are centered around sacred truths, such as "The Lord Our God the Lord is One," or "There is no god but God." These sacred truths are unverifiable, and unfalsifiable, but the faithful nevertheless accept them to be unquestionable. In doing so, like assemblies of the faithful since the dawn of language, they bind themselves together for protection or common action against the unbelievers and their lies.

From his study of the Maring, primitive agriculturalists of the New Guinea central highlands, Rappaport also recognized that ritual was an essential source of authority in an egalitarian society without headmen or ruling elites. It was by their attendance at ritual dances that the Maring would commit themselves to fight as their host's allies in the next war cycle. "It is plausible to argue that religious ritual played an important role in social and ecological regulation during a time in human history when the arbitrariness of social conventions was increasing but it was not yet possible for authorities, if they existed at all, to enforce compliance," he wrote.

Rappaport's ideas about the role of religion in early societies have been buttressed by a remarkable series of excavations in the Oaxaca valley of Mexico. The archaeologists Joyce Marcus and Kent V. Flannery traced the development of religion over a 7,000-year period as the people of the valley went through four stages of social development, from hunters and gatherers, to a settled egalitarian society, to a society ruled by an elite, and finally to an archaic state known as the Zapotec state. As the Oaxacan people's society evolved, so too did their form of religion.[210]

At the hunting and gathering stage, Joyce and Flannery found signs of a plain dance floor, its sides marked by stones. The dance floor, assuming it was used like those of modern hunter-gatherers, would have been the site of ritual dancing on ad hoc occasions when many different groups came together for initiations and courtship.

By 1500 BC the Oaxacans had developed strains of maize that allowed them to settle down and practice agriculture (the reverse of the sequence in the Near East, where settlement long preceded agriculture). At first their society was egalitarian, as it had been in their hunter-gatherer days, but their rituals became more formal. Marcus and Flannery have excavated four men's houses, all oriented in the same direction, which may have been determined by the sun's path at spring equinox. The orientation suggests that religious ceremonies were now held at fixed times, determined by astronomical events. The men's houses, to judge by practice in contemporary societies, may have been open only to men who had passed acceptability tests and been initiated into secret rituals.

By 1150 BC the third stage of society had began to emerge, with an elite who lived in large houses, wore jade-studded clothes and deformed their skulls in childhood as a sign of nobility. The men's houses were replaced with temples, also oriented in the same direction. Religious practice had become more elaborate, the archaeologists found, with ritual bloodletting, a symbolic self-sacrifice, and the cooking and eating of sacrificial victims.

The fourth stage of society, the Zapotec state, which was founded in 500 BC, was accompanied by a more complex form of religion. The temples now had rooms for a special caste of religious officers, the priests.

The advent of the priests marked the culmination of a steady trend in the evolution of Oaxacan ritual, its growing exclusivity. At the hunter-gatherer stage, the ritual dances were open to everyone. By the time of the men's houses, only initiated members of the public could participate in rituals, and by the stage of the Zapotec state, religion had come under the control of a special priestly caste.

What underlay this coevolution of religion with social structure? It seems that the important coordinating role of ritual in hunter-gatherer societies did not end when leaders and elites emerged in settled societies. Instead, the elites coopted the ritual practices as another mechanism of social control and as a means of justifying their privileged position. Making the religion more exclusionary gave the elites greater power to control the believers. To justify the ruler's position, new truths, also unverifiable and unfalsifiable, were added as subtexts to the religion's sacred postulates, such as "The chief has great *mana*," "Pharaoh is the living Horus," or "Henry is by the Grace of God King."

Rappaport believed that the conditions that enabled authorities to exercise civil power emerged only recently, and that for much of human existence rulers invoked sanctity as a principal source of their authority. Even archaic states were theocratic, at least to begin with. Modern states too, despite the ample civil power at their disposal, have not entirely dispensed with appeals to religious cohesion and authority. Even in a society like that of the United States, political allegiance is sealed with the declaration of "One nation under God."

Religion's other ancient role, that of protecting the community from freeloaders, can also been seen still at work in contemporary societies. Among ultra-Orthodox Jews in New York's diamond district, the level of trust is so high that multi-million-dollar deals can be sealed by a handshake. Islam is said to have spread through Africa as a facilitator of trade and trust.[211]

Trust and cohesiveness are nowhere more important than in wartime. Contemporary religions preach the virtues of peace in peacetime but in war the bishops are expected to bless the cannon, and official churches almost always support national military goals. "Religion is superbly serviceable to the purposes of warfare and economic exploitation," writes the biologist Edward O. Wilson, noting that it is "above all the process by which individuals are persuaded to subordinate their immediate self-interest to the interests of the group."[212]

Why does religion persist when its primary role, that of providing social cohesion, is now supplied by many other cultural and political institutions? While religion may no longer be socially necessary, it nevertheless fills a strong need for many people, and this may reflect the presence of genetic predisposition. Wilson, for one, believes that religion has a genetic basis, that its sources "are in fact hereditary, urged into birth through biases in mental development encoded in the genes."

Religion, language and reciprocity are three comparatively recent elements of the glue that holds human societies together. All seem to have emerged some 50,000 years ago. But a far more ancient adaptation for social cohesiveness, one that set human societies on a decisively different path from those of apes, was the formation of the pair bond. Much of human nature consists of the behaviors necessary to support the male-female bond and a man's willingness to protect his family in return for a woman's willingness to bear only his children.

The Privatization of Sex

Ape societies are driven by intrasexual competition, the rivalry between males for access to females. Although male chimpanzees form coalitions to seize power within the male hierarchy, these are shifting, ad hoc arrangements. A male chimp probably sees most other adult males as potential rivals, an attitude that limits the degree of cohesiveness in chimp society.

The human line of descent probably inherited the ape system of separate male and female hierarchies. But around 1.7 million years ago, the size difference between males and females started to diminish, according to the paleontological record. This shift in size is almost certainly a sign that competition between males had diminished because of the transition to the pair bond system.

The novel arrangement of pairing off males and females creates a whole new set of social calculations. Most males in the society now have a chance to reproduce since they possess socially endorsed mating access to at least one female. So each male has a much greater incentive to invest in cooperative activities, such as hunting or defense, that may benefit the society as a whole.

The pair bond takes much of the edge out of male-to-male aggression. It also requires that men trust one another more, and can have some confidence that those who go hunting won't be cuckolded by those who stay to defend the women.

For the females there is a trade-off. They must give up mating with all the most desirable males in the community and limit their reproductive potential to the genes of just one male. On the other hand they gain an implied guarantee of physical protection for themselves and their children, as well as some provisioning. In some foraging societies the men bring back meat but the staple foods are plants and small animals that are mostly gathered by the women. But a man's food gathering efforts were probably particularly helpful during the frequent periods that a woman was nursing and could find less food for herself.[213]

Like most of evolution's behavioral arrangements, pair-bonding was not a rigid prescription, one that dictated a one man, one woman nuclear family. Many human societies are polygamous, allowing men to have more than one wife if they can support more. Societies in special ecological conditions allow women to have more than one husband, such as in the high altitude agriculture of Tibet where a set of brothers may marry one wife and raise the children as a single family.

With the institution of pair bonds, sex became something conducted within families. It was presumably at this time that human societies developed a taboo against public sex, a custom that would bring chimp or bonobo societies to an almost complete standstill. The privatization of sex would help considerably in removing sex as a provocation of male rivalry.

But though the pair bond system alleviated discord between males, it raised new tensions between men and women. The asymmetry between the male and female roles in the family unit—a woman looks after the children, a man protects and supports her and her children—sets up an inevitable difference of reproductive interests and strategies. Men have evolved traits like sexual jealousy, for the sound reason that complaisant husbands are likely to pass on fewer of their genes.

Sexual infidelity poses very different kinds of risk for men and for women. For a woman, the threat posed by her husband having a mistress is not so much the sexual dalliance in itself but rather the possibility that he may switch his support to his paramour and away from his family. The withdrawal of support would reduce her reproductive fitness, as measured by the chance of raising her children successfully to maturity.

Serious as that danger may be to a woman's interests, the risk to a man of his wife's infidelity is considerably graver. For him, the threat is that he may not be the father of his wife's children. In evolutionary terms, a man who devotes his life to raising another man's children has seen his Darwinian fitness reduced to zero.

Men's fear of being deceived is not without basis, in a general sense, since a woman has a heavy incentive to seek another partner should her husband prove infertile, a not uncommon occurrence. Even if her husband is fertile, a woman might improve her reproductive success by having children with more than one partner. Ideally a woman will seek both support and good genes from her husband, but as long as that support is guaranteed her reproductive interests could in principle be improved by seeking better genes elsewhere. Many men may be willing to offer her this service, since they can greatly improve their reproductive success by having children with as many women as possible, especially if another man bears the cost of raising them.

It is no surprise, therefore, that men have gone to great lengths to secure exclusive access to women whose children they have undertaken to raise, with methods that range from foot-binding and genital mutilation to purdah, veils, chadors and an array of laws and customs restricting women's activities. However abhorrent the means, the motivation stems from the inherent vulnerability of male reproductive strategy: mother's baby, father's maybe.

How often do women conceive children with men who are not their husbands? Ornithologists used to rhapsodize about the marital fidelity of bird species that stayed pair-bonded for life. That was until the advent of protein-based and later DNA tests for assessing paternity. Despite the appearance of fidelity, extra-pair liaisons in the bird world turned out to be routine. The preeminent adulteress is an Australian bird, the Superb Fairywren, 76% of whose offspring are fathered by extra-pair copulations.[214]

Human geneticists testing people for heritable diseases quite frequently stumble across cases where the father of record cannot be the biological parent. Genetic counselors have a rule of thumb that these discrepancies, known delicately as nonpaternity cases, will range from 5 to 10% in an average American or British population. For the U.S. population as a whole, "The generic number used by us is 10 percent," said Bradley Popovich, vice president of the American College of Medical Genetics.[215]

The degree of nonpaternity that has come to light in the United States and Europe is particularly surprising in light of the control that women now exert over their reproductive behavior. Presumably many of the children involved in nonpaternity cases are not conceived by accident. The evident implication is that the woman's conception with a man other than her husband is in some cases deliberate.

That women in modern societies sometimes choose to conceive with alternative partners is a matter that bears on an issue of considerable debate among primatologists, that of whether the phenomenon of sperm competition occurs to any significant degree in people. In many species the female is inseminated by more than one male at the same time, and direct competition takes place in the female's reproductive tract between the sperm of rival males to fertilize the eggs. The female reaps the significant genetic benefit of having her eggs fertilized by the best of the competing sperm. Has evolution dispensed with this useful grading method in humans or does it apply in our species too? "Sperm competition is possible in *Homo sapiens*, though whether it has played a significant role during human evolution remains highly debatable," says Alan Dixson, an authority on primate reproduction.[216]

Geneticists have recently studied the DNA sequence of several genes involved in sperm production in three primate species, chimpanzees, gorillas and people. In chimpanzees, among whom sperm competition is fierce, the genes show signs of being under strong selective pressure. The pressure is much less fierce in the gorilla version of these genes, as would be expected given the silverback's exclusive access to his harem. In humans the genes have evolved at a rapid clip, faster than that of gorillas and equal to that of chimps. The human sperm genes are clearly under some kind of fierce evolutionary pressure, and sperm competition may be the reason.[217]

Sperm competition requires not just that a woman has more than one lover but that she has two within a rather short time of each other.[218] Some 4% of people in Britain are conceived under such competitive conditions, according to Robin Baker, a University of Manchester biologist.[219] This estimate receives some support from data on heteropaternity, a phenomenon that occurs in fraternal twins. Unlike identical twins, who arise from splitting of the same egg, fraternal twins result when the mother releases more than one egg in an ovulatory cycle. Heteropaternity refers to the circumstance in which each of two eggs is fertilized by different fathers. In Greek mythology the twins Castor and Pollux were the sons respectively of Tyndareus, a king of Sparta, and Zeus, who in the guise of a swan seduced Tyndareus's wife Leda. This may have been the first case of heteropaternity but it was by no means the last. Of fraternal twins born to white women in the United States, 1 in 400 pairs is estimated to have two fathers.[220] Among cases where paternity is disputed, 2.4% of cases have been found to be heteropaternal.[221]

For a woman to have a child extramaritally carries a serious risk—that the infant of an extramarital liaison may look suspiciously unlike the father of record, putting both its own life and its mother's at risk. This hazard would have created a strong selective pressure in favor of genes that prevented infants from looking too much like their parents. And indeed babies tend to have chubby faces with indistinct features that give them a rather generic appearance, sharply mitigating the risk that they will look like the wrong father.

To the extent they resemble anyone, babies would be expected to look as much like their mother as their father. But researchers have found that grandparents and others comment far more frequently on a baby's similarities to its father. Mothers tend to state that a baby resembles its father, and do so more often when the father is present, as if trying to assure him of his paternity. "Whether mothers do this consciously, knowing full well that the baby looks nothing like its dad, or whether they deceive themselves into thinking that the baby really does look like the father is unclear," say the authors of a textbook on evolutionary psychology.[222]

Because of the central significance of reproductive success, evolutionary psychologists have paid particular attention to human mating habits, exploring the signals that govern male and female choice of a mate, and the strategies that each pursues to accomplish its reproductive goals.

In studying the mating signals that the human psyche is genetically primed to assess, evolutionary psychologists have found that men in many different cultures of the world prefer women with a waist to hip ratio of 7 to 10. The male eye is probably attuned to these proportions because they signal a woman's fertility. Young women tend to put fat on their hips, breasts and buttocks whereas older women, and those who are pregnant, get thicker at the waist. "A relatively narrow waist means 'I'm female, I'm young, and I'm not pregnant,'" writes the evolutionary biologist Bobbi Low.[223] Symmetry of features, especially of the face, is another indicator of good genes; it requires a normal development in the womb and is thus a marker for general health. There are of course variations on the general theme. Among the !Kung, men are driven wild by a sizable protuberance of fat on a woman's buttocks, presumably a signal of being able to nourish a child in difficult environments.

Because reproduction is a greater risk and investment for women than for men, according to the biologist Robert Trivers, it follows that women will be more choosy about their partners than men are; and because women are more selective, men will find themselves being more competitive with each other for women's favors. A woman looks for indicators not just of good health in a man but also of commitment to look after her and her family. This is a matter partly of emotional commitment, which women assess with care, and also of wealth or the ability to acquire it, as may be indicated by a man's social status.

Surveys conducted over many years have consistently indicated that American women care more about a partner's wealth than men do. The evolutionary psychologist David Buss expanded this survey to 10,000 individuals in 37 world cultures and found the same pattern—that women placed more value than did men on a partner's financial prospects.[224] Women in almost all cultures prefer men of high status, presumably because this is likely to be correlated with wealth. Women consistently prefer men who are slightly older, for reasons that are not obvious. The preference could be a holdover, Buss suggests, from hunter-gatherer days when older men, at least through their twenties and thirties, were stronger and better able to offer physical protection to their family.[225] Perhaps for the same reason, women consistently prefer tall men to short.

If fitness indicators for health and fertility are useful, wouldn't indicators for mental ability be even more useful in choosing a partner for a long-term relationship? The evolutionary psychologist Geoffrey Miller has advanced the striking theory that such indicators do exist, but they are familiar under names that give no clue to their biological function. The indicators of mental fitness, in his view, include both cultural activities such as art, music, dance and literature and moral qualities such as kindness.[226]

Evolutionary biologists have gained considerable insight into what makes fitness indicators true signals, and why they must be qualities that vary in a population. Fitness indicators, and the behavioral preferences for them, are brought about by sexual selection, a form of natural selection but one that works through mating success rather than physical survival. The mechanism of sexual selection was first recognized by Darwin, who had long been puzzled why the males of many species are heavily ornamented, with conspicuous horns or antlers or feathers. These baroque decorations seemed to contribute nothing to survival, posing an apparent challenge to Darwin's idea that the fittest survive. The solution he proposed in *The Descent of Man* was female choice: peahens for some reason preferred peacocks with gaudy tails, who got to sire more offspring, including sons with gaudy tails and daughters with a taste for them. These male adornments were therefore a worthwhile handicap to their owners because they assisted toward evolution's bottom line of getting more genes into the next generation.

That still left the question of how these male embellishments evolved in the first place. Darwin's theory of sexual selection was largely ignored for a century—his contemporaries placed no credence in the idea that female choice could be a major evolutionary force—and it was not until the 1970s that biologists started to develop the theory. One insight was that male ornaments like long plumage were hard to grow and therefore served as an overall indicator of good genes. But if long red tail feathers, say, were the key to male reproductive success, soon every male would be wearing them and they would lose their utility in helping females choose between males of different quality.

The evolutionary biologist Amotz Zahavi realized that sexual signals of one's health and fitness, if they were to be true and reliable, had to be so costly as to constitute a serious handicap for the displayer. Weak peacocks grow unappetizing tails and only the strongest can grow really beautiful ones. It's that spectrum of ability that provides peahens with a basis for choice. Biologists call such a trait heritable—the quality of the tail varies from one individual to another and part of the variation is caused by the genes.

The physical features that have evolved as fitness indicators in the human mating dance—symmetry of features, fine skin, a shapely body—are known by another name: beauty. People find these features attractive not through some arbitrary criterion or dictate of fashion but because the male and female minds have evolved to look for and appreciate such qualities in a potential mate.

Zahavi's costly signaling theory explains, sad to say, why it is impossible for everyone to be equally beautiful or handsome. Since beauty serves as a fitness indicator, it needs to vary from one individual to another. If everyone were equally beautiful, beauty would have no value as a criterion for sexual selection.

The privatization of sex that began 1.7 million years ago did not bring an end to all competition between males for females. But it was a major step in reducing human aggressiveness within societies. And it was followed, many thousands of years later, by a serious evolutionary reduction in the level of aggression between societies.

The Domestication of People

The evidence that human tribes have become less passionately hostile to each other lies in a worldwide thinning, or gracilization, of the human skull that took place during the Upper Paleolithic era. The fossils of early modern humans are both large and very robust, or thick boned. But these generic early modern skulls started to change around 40,000 years ago. In each region of the world they follow an independent, but largely parallel course, as if similar genetic changes are occurring independently in each population. "Cranial size reduction and gracilization may have been homoplasic [arising by independent evolution] in most populations," writes the physical anthropologist Marta Mirazón Lahr.[227]

The gracilization occurred at different rates in different regions but all followed a common trend, except for two populations at the extremities of the human diaspora, Australian aborigines and Fuegians at the tip of South America. The Australian skulls became smaller like the rest, but retained their robusticity, presumably as a result of independent evolution. The Fuegians seem to be a case of genetic drift—a small isolated population developing its own special characteristics.[228] Gracilization is farthest advanced in sub-Saharan Africans and Asians, with Europeans still in some instances showing large size and robusticity.[229]

What caused the gracilization of human skulls and the shrinking of human skulls and teeth all over the world? This is a large and complex issue, not least because a very similar downsizing affected the sheep, goats and other animals that were domesticated in the Neolithic era after the advent of agriculture. Researchers have attributed the smaller size of domesticated animals (compared with their wild forebears) to such facts as different diet, less physical activity, and a relaxation of the selective pressure favoring larger males under the conditions of captive breeding.[230]

Lesser physical activity and the dietary changes brought about by agriculture have also been suggested as the reason why humans became lighter-boned and smaller. But the explanation must be sought elsewhere because it is clear that the gracilization of humans started well before the beginning of agriculture, and around the time of the earliest settlements some 15,000 years ago. The Natufians, the first settlers of the Near East, already had more gracile features, shorter stature and smaller teeth.[231]

Lahr believes that the more gracile features appearing in human skulls of the Upper Paleolithic "have a strong genetic basis," but her study is purely descriptive and she offers no explanation for the forces that might have driven the genetic change. The primatologist Richard Wrangham, however, has provided an intriguing insight into gracilization.

His argument goes as follows. Consider first the bonobos, who are much more peaceful and playful than chimpanzees. Their skulls look like those of juvenile chimpanzees, just as their behavior is more juvenile than that of chimpanzees. This kind of change is called pedomorphic—meaning a trend toward the juvenile form—in reference to the evolutionary process of developing a new species by truncating the fully mature development of the ancestral species. Bonobos presumably found themselves in an environment where aggression was less beneficial, and so evolution kept selecting individuals whose development was completed before the arrival of the aggressive traits typical of adult males.

Pedomorphic evolution is familiar to biologists in another context, that of domestication. Comparing dogs with wolves, the dog's skull and teeth are smaller and its skull looks like that of a juvenile wolf. The same process occurred when Dmitri Belyaev, in the experiment already discussed, set out to domesticate silver foxes. Belyaev selected foxes solely for tameness, but a whole set of other traits appeared in his animals along with the tolerance of people, including the white marks on the coat, curly hair, and smaller skulls and brains.

Viewed in this context, the gracilization of the human skull looks very much like one of those changes that come along for the ride when a species is undergoing pedomorphosis or domestication. Gracilization, Wrangham believes, occurred because early modern humans were becoming tamer.

And who, exactly, was domesticating them? The answer is obvious: people were domesticating themselves. In each society the violent and aggressive males somehow ended up with a lesser chance of breeding. This process started some 50,000 years ago, and, in Wrangham's view, it is still in full spate. "I think that current evidence is that we're in the middle of an evolutionary event in which tooth size is falling, jaw size is falling, and it's quite reasonable to imagine that we're continuing to tame ourselves. . . . This puts humans in a picture of now undergoing a process of becoming increasingly a peaceful form of a more aggressive ancestor."[232]

With tamer people, the path was now set for larger and more complex societies, ones that would transcend the limited horizons of the hunter-gatherer band.

The Progression of Human Society

The vocabulary of evolutionary biology does not include the word progress, for evolution has no goal toward which progress might be made. But in the case of human evolution, this exclusion may not be entirely justified. People, after all, make choices. If those choices shape a society for generation after generation, and if they permit individuals of a certain character to have more children and propagate their genes, then the overall nature of society may come to be shaped, in part, by human choice. If the character in question is a tendency to cooperate with others, then such a society would become more cohesive internally and more conciliatory in its relations with neighbors. Other societies might become more aggressive in character, or more paranoid, or more adventurous. Yanomamo society, given that the *unokais* have more children, has surely been positioned to become more aggressive. But overall, despite many setbacks and reversions, human societies have made vast gains in peacefulness, complexity and cohesion in the last 15,000 years.

It is often assumed that evolution works too slowly for any significant change in human nature to have occurred within the last 10,000 or even 50,000 years. But this assumption is incorrect. The development of new brain gene alleles 37,000 and 6,000 years ago, and of lactose tolerance 5,000 years ago, have already been mentioned; several other instances of recent human evolution are cited in chapter 12. There is no reason to suppose that human nature ceased to evolve at some finishing post in the distant past or to assume, as do some evolutionary psychologists, that people are struggling to function in modern societies with Stone Age minds. Genomes adapt to current circumstances or perish; the human genome is unlikely to be an exception.

Human societies have progressed through several major transitions in the last 15,000 years, and it may well be that these transformations were accompanied by evolutionary as well as cultural changes. It was only after people had become less violent that they were able to abandon the nomadic life of hunting and gathering that they had followed for the last 5 million years, and began to settle down. The first settled societies appeared in the Near East some 15,000 years ago. Though they were probably egalitarian at first, they soon developed a hierarchical form, with elites, leaders and specialization of roles.

Once settlement began, human societies became larger and more complex, presenting a new set of environments for people to adjust to. Societies come in many forms, and each may have punished or rewarded different character traits. The anthropologists Allen Johnson and Timothy Earle have traced the emergence of human societies of various levels of complexity, arguing that each is a response to the environmental problems it had to tackle, notably those of food production, surpluses, defense and trade. They distinguish three broad levels of complexity—family-based societies, local groups and regional polities.[233] Each of these major cultural transitions could well have prompted changes in social behavior and these, though Johnson and Earle make no such suggestion, could have become genetically embedded as the individuals who best adapted to each new social stage left more children.

Hunter-gatherer societies, Johnson and Earle say, were based on fairly autonomous family groups, though with a degree of organization that extends beyond the family. To spread the risk of catching nothing, hunters like the !Kung have firm rules for distributing the meat from a kill beyond the hunter's immediate family. A large animal may have more meat than a single family can consume, so sharing it buys entitlement to a reciprocal gift in future.

Two themes already apparent in foraging societies—reciprocity and leadership—emerged more strongly in settled societies. Settled societies, in the Johnson-Earle analysis, needed assurance of food supply. But instead of sharing on an ad hoc basis, as foragers do, they had another option, that of generating and storing surpluses.

Surpluses, largely unknown to hunter-gatherers, were of critical importance to settled societies. The surpluses had to be stored, protected and distributed, activities that required a greater level of social organization than the loose associations of a family-based foraging group. Local groups emerged, like a Yanomamo village, in which there was a headman, though with few powers beyond those of personal persuasion. Religious ceremonies played a leading role in integrating group activities.

Surpluses also generated items that could be traded. The increasing complexity of managing a local group's trade, defense and investment (such as in fishing weirs or irrigation) required stronger leadership. Eventually chiefs emerged, along with specialists and elites. These leaders integrated village-size communities into a regional economy by managing long distance trade and spreading the risks of food production beyond the family level.

The ground had then been laid, Johnson and Earle suggest, for the association of local groups into a larger society. Continuing intensity of economic activity led to the emergence of the first states, known as archaic states. In Japan, for example, people lived as hunter-gatherers until around 250 BC when the cultivation of dry rice was introduced. Foraging and dry rice farming existed side by side until AD 300 when wet rice began to be cultivated. This required large scale irrigation, and at the same period the first chiefdoms and archaic states emerged.

Archaic states have existed only in the last 5,000 years. During Neolithic times, Johnson and Earle estimate, there were probably more than 100,000 independent political units of the family-based or local group level of organization. But at all levels of the social organization, from hunter-gatherers to archaic states, the goal was the same, that of organizing resources in a way that benefited the reproductive strategies of its members.

In the emergence of these early human states, two strong forces were at work, and still shape relations between states in the contemporary world. One is the need for defense, the other the dependence on trade. Both of these state behaviors spring from the deepest wells of human nature, the contrary instincts for aggression and reciprocity. Though war gets more space in the history books, it is the conciliatory arts of trade and exchange that have prevailed in the long run. According to the World Health Organization, only 0.3% of deaths in 2002 were caused by war.[234]

Our bones are more gracile than those of our Upper Paleolithic ancestors, our personalities less aggressive, our societies more trusting and cohesive. An element of human choice, a preference for negotiation over annihilation, has perhaps been injected into the genome. And that might explain why there is an inescapable sense of progress about human evolution over the last 50,000 years: human choice has imposed a direction on the blind forces that hitherto have shaped evolution's random walk.

In parallel with human social evolution, the human physical form continued to evolve. Because the human population was dispersed across different continents, between which distance and hostility allowed little gene flow, the people on each continent followed independent evolutionary paths. It was these independent trajectories that led over the generations to the emergence of a variety of human races.

9

RACE

Although the existing races of man differ in many respects, as in
colour, hair, shape of skull, proportions of the body, &c., yet if their
whole structure be taken into consideration they are found to resem-
ble each other closely in a multitude of points. Many of these are of so
unimportant or of so singular a nature, that it is extremely improb-
able that they should have been independently acquired by aborigi-
nally distinct species or races. The same remark holds good with equal
or greater force with respect to the numerous points of mental simi-
larity between the most distinct races of man. The American aborig-
ines, Negroes and Europeans are as different from each other in mind
as any three races that can be named; yet I was incessantly struck,
whilst living with the Fuegians on board the Beagle, with the many
little traits of character, shewing how similar their minds were to ours;
and so it was with a full-blooded negro with whom I happened once to
be intimate.

CHARLES DARWIN, THE DESCENT OF MAN

AFTER THE ANCESTRAL PEOPLE had dispersed from their homeland in northeast Africa, there was no longer a single human population but many. Across the far-flung reaches of the globe, human
evolution continued independently. Over the course of many generations
the peoples of each continent emerged as different races.

Such an outcome is not so surprising. An array of influences would have
pushed each population along a separate evolutionary path. And the one
force that could have kept the population the same—a thorough mixing of
genes, through intermarriage—could no longer operate once people lived
vast distances from each other, and probably in warring tribes who killed
as spies anyone found traveling through their territory. The genealogies of
the Y chromosome and mitochondrial DNA, whose major branches are still

largely confined to different continents, are evidence that throughout the world people have tended overwhelmingly to live, marry and die in the places they were born, at least until modern times.

The genetic differentiation of the human population, into races and ethnicities within races, has long been a matter of both controversy and ignorance. Because of the many evils that racism has caused, from discrimination to genocide, researchers have generally sought to minimize the existence of race. Many social scientists even assert that race is a social concept without biological basis.

Race is not well understood, in part because it has not been regarded as a fit subject for academic study. In many respects, this has been a prudent position. The matter of race seemed of no great scientific interest, was inherently divisive, and had been seriously polluted by a history of racial classifications designed with an agenda of proving one race superior to another.

But two valid scientific reasons for considering the question of race have begun to emerge, and at the same time technical advances in sequencing DNA have at last made it possible to study the still somewhat mysterious nature of race on a scientific basis.

One valid reason for reconsidering race is historical; people of different races may hold in their genetics essential clues to human history since the fragmentation of the ancestral human population 50,000 years ago. Races presumably developed in part in response to the pressures experienced by each population, and the genetic changes involved in race may allow those pressures to be identified. The different branches of the human family have their own histories, which cannot be explored or told until the branches are recognized and their genetics examined.

A second and more practical reason for defining race is medical. Many diseases have a genetic component, which often varies with race or ethnicity. Hemochromatosis, a genetic condition thought to have been spread by the Vikings, affects mostly Europeans. The Pima Indians are particularly susceptible to diabetes, Pacific Islanders to obesity. Crohn's disease occurs in both Europeans and Japanese but the three genetic variants known to be the cause of the disease in Europeans are not found in Japan, where presumably a different mutation leads to the same symptoms.

People of different races may also differ in their response to drugs. This is sometimes because the enzymes that break down the drugs are being lost at different rates in different races. (The enzymes' original role was to break down the natural toxins in wild plant foods; since they are no longer needed for that purpose, they are being randomly inactivated by mutations that natural selection no longer sweeps away.) People may also possess different versions of the protein on which a drug is meant to act. The heart drug enalapril reduces blood pressure and the risk of being hospitalized for heart failure in white patients but has little effect in blacks.[235]

Another drug to which races respond very differently is the new heart failure treatment known as BiDil, a combination of two existing drugs invented by Jay N. Cohn, a cardiologist at the University of Minnesota. On its first trial, in a general population, BiDil didn't appear to be particularly effective. But Cohn noticed on further analysis that it seemed to have done well in a subset of the population, who turned out to be African Americans. In a new trial, undertaken in African Americans alone, BiDil proved to work so well that the trial had to be stopped in order that the drug could be offered to patients in the comparison group who were not receiving it.[236] (BiDil may be effective in people of African ancestry because, as a way of retaining salt in hot climates, they have genetically low levels of a chemical signal that BiDil enhances.)

The emergence of a genetically different pattern of disease and drug response in various populations has touched off a vexed argument about race based medicine. Some physicians contend that consideration of a patient's race is not or should not be a proper part of medicine. But some geneticists differ strongly, saying that the human genome sequence has now made it possible to tailor diagnosis and treatment to each population's special needs, and that it would be folly to ignore racial differences if, as in the case of BiDil, race is the key to discovering effective therapies.

Neil Risch, an eminent geneticist now at the University of California, San Francisco, was the first to say in print that the emerging view of human population structure had major points of correspondence with the public conception of race. Risch's article was sparked by his irritation at the sociologists' race-is-not-biological dogma surfacing in, of all places, the *New England Journal of Medicine*, a leading journal of medical research. "Race is a social construct, not a scientific classification," declared an editorial by Robert S. Schwartz, the journal's deputy editor.[237] Since race is "biologically meaningless," Schwartz argued, it should not play any part in a physician's work. A similar editorial, though less absolutist, appeared in the journal *Nature Genetics*.[238]

Much of this discussion, Risch wrote in rebuttal of the two editorials, "does not derive from an objective scientific perspective." (In the determinedly dull parlance of the scientific literature, these are fighting words.) Numerous genetic studies of the human population have found that differences are greatest between continents. These studies, he said, "have recapitulated the classical definition of races based on continental ancestry." Updating those definitions, Risch and his colleagues suggested that racial groups should be defined on the basis of continent of origin, with ethnicity being used to describe smaller subdivisions within races.

The five continent-based races, in Risch's view, are as follows:[239]

Africans are those whose primary ancestry is in sub-Saharan Africa. This includes African Americans and Afro-Caribbeans.

Caucasians are people of western Eurasia—Europeans, Middle Easterners, North Africans and those of the Indian subcontinent (India and Pakistan).

Asians are people of eastern Eurasia (China, Japan, Indochina, the Philippines and Siberia).

Pacific Islanders are Australian aborigines and people of New Guinea, Melanesia and Micronesia.

Native Americans are the original inhabitants of North and South America.

Within each continental race there are gradations of skin color, from the light-skinned Khoisan speakers of southern Africa to the darker-skinned Bantu speakers of western and central Africa, from the lighter-skinned Scandinavians to the darker-skinned peoples of southern India. Skin color is therefore an ambiguous indicator of continental race.

At the boundaries of these continental divisions are several groups formed by intermarriage between the two neighboring races, a condition for which geneticists use the term "admixture." Ethiopians and Somalis, for instance, are an admixture of Caucasians and Africans. "The existence of such intermediate groups should not, however, overshadow the fact that the greatest genetic structure that exists in the human population occurs at the racial level," Risch says.

In the United States there are several populations formed by intermarriage between members of two racial groups. African Americans, largely as a result of slavery, have a share of Caucasian genes that ranges from 12% to 23% in various populations, with an average of about 17%. "Despite the admixture, African Americans remain a largely African group, reflecting primarily their African origins from a genetic perspective," Risch says.

Another group of admixed populations is counted by the U.S. Census Bureau as Hispanic although Hispanic is a linguistic, not a racial, category. Hispanics vary in their admixture in different parts of the country. In the southwestern United States, Hispanics are mostly Mexican Americans, whose ancestry is 39% Native American, 58% Caucasian, and 3% African, according to one recent estimate. East coast Hispanics come mostly from the Caribbean and have a larger proportion of African genes.

The United States is often referred to as a melting pot of races but the rate of mixing is slower than might be assumed. Figures from the 2000 U.S. census indicate that U.S. citizens do not marry each other at random. Racial endogamy (marrying within the racial group) is the rule: 97.6% of respondents reported themselves to be of one race; only 2.4% said they were of more than one race, presumably having parents of different races. Some 75% of Americans declared themselves to be white, that is, Caucasian; 12.3% said they were black or African American; 3.6% were Asian, 1% Native American, and 5.5% of other races.

Differentiation of the
Ancestral Human Population

These continental groups reflect the leading roles of geography and endogamy in shaping human races. As long as everyone intermarries, as would doubtless have been the case in the ancestral human population, there is a single genetic pool. New diversity—that is, new alternative versions of genes—accumulates through mutation, and old diversity is eliminated by drift, but these changes occur within a common pool. Any substantial bar to

intermarriage, however, whether a mountain range or a religious ban on marrying outsiders, will set up two genetic pools. Since mutation and drift are
both random processes, the changes in the two pools will now take place independently. From that point on, the two populations may follow different
evolutionary paths. Migration between the two will sharply reduce genetic
difference; time and distance will increase it.

The starting point for the emergence of human races would have been
the dispersal, within Africa, from the ancestral homeland some 50,000 years
ago. Before people left for the world beyond, the human population in Africa
had apparently fragmented, doubtless by geographical distance, into several
different populations. As already noted, those who left Africa belonged to
just one of these populations, those descended from the L3 branch of the
mitochondrial DNA tree. They carried away in their genes only a subset of
the African genetic diversity, meaning only some of the alleles of each gene.
That fact alone set them on a potentially different evolutionary path.

The emigrants eventually spread out over the rest of the globe and
themselves fragmented into many even smaller populations. The smaller a
population, the greater is the force of genetic drift, which reduces the number of available alleles. Without interbreeding to keep the human gene pool
mixed, the populations of each continent or region would over time have become more distinct and less like the others.

The importance of drift in differentiating a static population has recently been recognized in the population of Iceland.[240] As mentioned earlier, even though the island has been settled for only 1,000 years, the people
in each region have become sufficiently different genetically that by sampling Icelanders' genome in just 40 different places it is possible to tell
which of 11 regions of the island they come from. In the rest of the world,
with some 50 times longer for genetic forces to act, and many severe impediments to movement, a much greater degree of differentiation would be expected.

Besides drift, another differentiating force on the world's separate human populations would have been natural selection. Selection may have
pressed particularly hard on the people who left the African homeland, since
they would have had to adapt to radically new diet, terrain and climates. A
particularly striking example of selection is a recently discovered gene vari-

ant that causes pale skin in Caucasians. Almost all African and Asians have the same, ancient form of the gene, which is known at present as SLC24A5. Some 99% or more of Europeans have a new version, that must have arisen after Caucasians and East Asians had become separate populations. The new version presumably became almost universal among Caucasians because the pale skin it conferred was of overwhelming advantage, whether for reasons of health or sexual attractiveness or both. A different gene, yet to be discovered, must give East Asians their pale skin.[241]

As Darwin suggested, sexual selection, the partly capricious taste of women and men for partners of a certain type, as well as competition between men, may have been a strong selective force, and one that acted somewhat independently in each human population. Disease has certainly influenced the human genome as people in different regions responded to local diseases like malaria. Warfare, an unremitting pressure, surely played a major role in shaping populations. And another powerful molder of human populations would have been climate, especially the adaptations necessary for living in northern latitudes and the violent climatic swings of the late Pleistocene.

Given all these evolutionary forces at work, it is not so surprising that the widely dispersed human populations in various continents acquired their own distinctive variations on the general human theme. This genetic-geographical difference is reflected in the familiar trees drawn on the basis of mitochondrial DNA or the Y chromosome, and on several other kinds of genetic elements. Risch cited some of these studies as proof of the division of the human population into continent based races.

A few months after Risch's article of 2002, a more comprehensive study by Marcus Feldman of Stanford University reached a very similar conclusion. Instead of examining just a few markers, or sites on the DNA, as many previous studies had done, Feldman and his colleagues looked at 377 sites throughout the genome, a larger and more representative sample. This was done for each of 1,000 people from 52 populations around the world. A computer was then instructed to group the individuals, based on their DNA differences at the 377 sites, into clusters. They fell naturally into 5 clusters, corresponding to their five continents of origin—Africa, western Eurasia (Europe, the Middle East, the Indian subcontinent), East Asia, Oceania and the Americas.[242]

Feldman and his colleagues did not use the word "race" in their article, referring instead to "structure" and "self-reported population ancestry" (meaning a person's own identification of their race), but he acknowledged in an interview that the finding essentially confirmed the popular conception of race. "Neil's article was theoretical and this is the data that backs up what he said," Feldman commented in reference to Risch's study.[243]

Identifying Race by DNA

A consequence of the Risch and Feldman studies is that they provide, for the first time, an objective way of ascertaining an individual's race. Most previous systems of race classification, with the principal exception of modern craniometry, have been based on characteristics like skin color, which vary in an unsystematic way, and were often designed with a malign agenda such as demonstrating one race's alleged superiority to others. Not only does the genetic definition of race have no such agenda, but it has nothing directly to do with any physical attribute.

The reason is that the genetic markers used to identify race are not part of the genes or their control regions, so far as is known, and therefore play no part in the physical appearance or behavior of an individual. Presumably they are indirectly correlated with genes that do control the body's physical makeup, but the connection is indirect and at present unknown.

The DNA markers analyzed by the Feldman team are of the same type as is used in the DNA fingerprinting of forensic cases. At various sites on the human genome the sequence of DNA units goes into a sort of stutter, known as a short tandem repeat because a few units of DNA are repeated several times over, as in AC-AC-AC-AC-AC. For some reason, these stutters tend to confuse the cell's DNA copying apparatus, which every dozen or so generations may accidentally either add or delete a repeat. The exact number of repeats at a given site is therefore quite variable from one person to another, and so can be made the basis of systems for identifying populations or individuals.[244]

Only some 3% of the DNA in the genome is devoted to genes; the rest of the DNA is mostly yards of filler material. The short tandem repeats are part of the filler material so do not affect a person's physical makeup. But some repeats lie close to genes, some of which have evolved in different ways in the various races. By selecting the right repeats, geneticists can find ones that are quite diagnostic of race, even though at present they have little idea which genes it is that give people of different races their different appearance.

Risch calculated that if the sites with short tandem repeats were chosen entirely at random, analysis of about 100 sites should suffice to say which of the five major races a person comes from. But as few as 30 sites would be enough if the sites were specially chosen so as to be diagnostic of race. Many hundreds of markers would be needed to distinguish, within a race, between two populations or ethnicities, Risch estimated.

Sets of these sites, known as Ancestry Informative Markers, can be used to identify not just an individual's race but the racial origin of individual sections of a person's genome. A company called DNAPrint Genomics has already started offering a test to assess people's continent of origin and, if of mixed race, the proportions of ancestry due to different races.[245] The test is based on a set of markers identified by Mark Shriver, a geneticist at Pennsylvania State University. It has already proved useful in police inquiries by identifying the race of a suspected serial killer from tissue collected at a crime scene. In June 2003 police believed that a serial killer in Louisiana was white, but were informed otherwise by DNAPrint Genomics, whose test showed the killer's ancestry was 85% African and 15% American Indian; they then arrested a suspect who was black.[246] The reliability of the test has not yet been established, but if it helps police identify a suspect, the suspect's DNA can then be compared with the crime scene DNA in the usual way.

Feldman and his colleagues say they needed varying numbers of markers—in this case sites with tandem repeats—to identify a person's continent of origin, depending on the genetic variability of the race in question. Native Americans could be assigned to their continent of origin with just 100 markers, whereas almost all 377 markers were required to identify Middle Easterners. This is because Native Americans are all descended from their Siberian founders whereas Middle Easterners are a more complex genetic blend; they are mostly Caucasian but some, like the Bedouin, have an African contribution.

Feldman's method gives a glimpse of how deeply genetic markers may be able to reach into population history. The computer program used to sort the genome samples into continental clusters could also split an individual's genome into different parts if the person was of mixed ancestry. People from the Hazara and Uighur of Central Asia, long a crossroads between east and west, emerged with genomes roughly half Caucasian and half East Asian in origin. The Surui, a fairly isolated people of Brazil, have genomes that are entirely American (in terms of the computer program's 5 racial clusters), whereas Mayan genomes are American with a strong dash of European and East Asian admixture.

With extra markers, and ones chosen to be more diagnostic of geographical origins, it should be possible to explore a population's ancestry and history in a much more detailed way. For geneticists, the essence of race is not politics but history: race defines through which branch of the human family tree people trace their descent.

Scientific Attitudes to Race

Researchers' attitude to race has swung through a wide arc in the last century and the new view developed by the work of Risch, Feldman and others has probably not yet become the consensus view.

In the nineteenth century, as European explorers became acquainted with the peoples of other continents who seemed so different from themselves, a serious debate arose as to whether these strange foreigners should be considered as belonging to separate species. Darwin, with his usual unerring insight, rejected the idea in his 1871 book *The Descent of Man*, arguing there was only one human species, though divided into subspecies or races. The one species must have had a single origin, which Darwin presciently placed in Africa, the continent with the greatest diversity of great apes. That human population, in his view, was later fragmented into different races by geographical isolation, followed by a differentiation that in Darwin's view was principally driven by sexual selection, the preference by women for men of a certain type.

Since Darwin's time, greater awareness has developed of the dangers of race in light of the many harms and injustices committed by people of one race against those of another. Many academic researchers, including geneticists, have sought to minimize the extent of biological variation within the human family. What is still one of the most influential positions on race is a statement made in 1972 by the geneticist Richard Lewontin.

Lewontin measured a property of proteins (DNA sequencing was not then available) taken from people of different races, and computed a standard measure of variation known as Wright's fixation index or F_{ST}. The idea is to measure some character that varies in members of a population and assess how much of the variation arises because two subpopulations differ from each other in that character. The index, in other words, reflects how much of the variation is general and how much is specific to the subpopulations.

Lewontin's value for F_{ST} came out at 6.3%, meaning that of all the variability in the human population, at least as reflected in the 17 proteins he had measured, only 6.3% lay between races, while a further 8.3% was found to lie between the ethnic groups within races. "Of all human variation, 85% is between individual people within a nation or tribe," Lewontin concluded.[247]

This finding is perfectly in line with the expectation that most of the genetic variation in each race would be the same as that of the ancestral human gene pool from which it was drawn. But the question then arose as to whether the extent of the difference between races was large or small. Lewontin argued that the difference was so trivial that racial classification had no genetic significance or justification.

Many biologists have chosen to go along with his interpretation, and this position has been followed, even taken to extremes, by the major social science organizations in the United States. According to the American Sociological Association, race apparently does not even have a biological foundation, since it is a "social construct." The association's official statement on race warns that "Although racial categories are legitimate subjects of empirical sociological investigation, it is important to recognize the danger of contributing to the popular conception of race as biological."[248]

The American Anthropological Association also dismisses the idea that biological differences can be recognized between races: "In the United States both scholars and the general public have been conditioned to viewing human races as natural and separate divisions within the human species based on visible physical differences," the AAA statement says. But since physical traits vary smoothly across the globe, and are not correlated with one another, "these facts render any attempt to establish lines of division among biological populations both arbitrary and subjective."[249]

But people can now be objectively assigned to their continent of origin, in other words to their race, by genetic markers such as those used by Feldman. And Lewontin's characterization of the differences he had found as trivial was as much a political as a scientific opinion. The degree of differentiation he had measured in the human population was similar to other estimates that put the value of global F_{ST} as between 10 and 15%. Sewall Wright, one of the three founders of population genetics and the inventor of the F_{ST} measure, commented that "if racial differences this large were seen in another species, they would be called subspecies."[250] Wright specified that an F_{ST} of 5 to 15% in any population of organisms constituted "moderate" genetic differentiation, and 15 to 25% should be considered "great" genetic differentiation.[251]

But whether an F_{ST} of 10 to 15% represents a large or small degree of differentiation is probably not very relevant to the question of whether or not human races can be identified. The reason is that such measures of F_{ST} have captured something called neutral variation, which is probably not the kind that underlies most racial differences.

Neutral variation refers to mutations that don't affect the organism one way or another. Evolution doesn't care about such changes, and the frequencies of such alleles in a population will vary randomly under genetic drift. Most common variation is neutral, and most measures of F_{ST} are likely to sample common—that is, neutral—variation.

But evolution pays great heed to mutations that make a significant change to a gene and its protein. If the change is adverse, the mutations are ruthlessly eliminated because the affected individual either dies or fails to reproduce. If the change enhances the individual's reproductive success, the mutation is selected for and becomes commoner in a population. These two kinds of selection, one negative and the other positive, are the two faces of natural selection. Biologists do not yet understand what genes need to be changed to make one species into two subspecies. However, it seems that the alleles involved in differentiating the human population are likely to be of the selected kind, not the neutral kind.

Versions of the two brain genes that evolved within the last 40,000 years show just this pattern. As mentioned in chapter 5, an allele of one, known as microcephalin, appeared some 37,000 years ago and is now widespread among Caucasians and East Asians but is much less common in sub-Saharan Africans. The F_{ST} for this allele between sub-Saharans and the others is 48% or 0.48, "which indicates strong differentiation and is significantly higher than the genome average of 0.12," writes Bruce Lahn of the University of Chicago, who discovered the allele.[252]

A new version of another gene, ASPM, arose some 6,000 years ago in Caucasians, 44% of whom now carry this allele. The allele is less common among East Asians and rare to nonexistent in sub-Saharan Africans. The F_{ST} for the allele between Caucasians and everyone else is 0.29.[253] There doubtless exist alleles of other brain-related genes, yet to be discovered, that are more common in East Asians or sub-Saharans and rare among Caucasians.

Most of the diversity in human skin color, for example, exists between populations, not within them. The F_{ST} for skin color is 88%, according to a study by John Relethford of the State University of New York at Oneonta.[254] Skin color is heavily correlated with latitude as well as race, but clearly does not follow the pattern of the neutral genes. Henry Harpending and Alan Rogers suggest that "other visible traits that most humans notice are more like skin color than they are like neutral traits"—in other words that most of the physical characteristics on which people judge a person's race are likely to be selected, just as would be expected if sexual selection has been the major force differentiating the human population.[255]

It is a few selected genes, not the many neutral ones, that may account for the differences between continental races. Substantial evidence for this idea has now emerged from a genomewide survey by Jonathan Pritchard of the University of Chicago. Devising a test to identify genes under recent selective pressure, he found roughly 200 such genes in Africans, in East Asians, and in Europeans. Each race's set of selected genes overlapped very little with those of other races, just as would be expected if the populations on each continent had adapted independently to evolutionary pressures.[256]

Genotype and Phenotype among Races

Even if geneticists can see a difference between races at the level of DNA, what practical difference does that make at the level of the physical person?

Biologists make a useful distinction between the genotype and phenotype of an organism. The genotype is simply the genome or hereditary information; phenotype is the physical creature that is generated from the genotype.

Because so much of the genome is nonworking DNA that does not code for genes, it is possible for genotype to vary without causing much change in phenotype. The Icelanders whose genotype can be matched to the island's 11 regions probably don't look any different from each other (although a survey is under way to see if they have a slightly different pattern of disease).

The work of Risch and Feldman showing that people can be genetically assigned to their continent of origin—that is, race—is based on genotype, and does not in itself indicate how much people of different racial genotypes might differ in phenotype. However, the sites on the genome that they examined move around during recombination (the shuffling of blocks of DNA that occurs between generations) in the same way that genes do. So some of these sites, especially those that lie close to genes, will be proxies for the genes themselves.

Thus the fact that people can be assigned to racial groups based on sampling just a few hundred sites in their genome suggests that quite a large number of genes may also vary between races and that so may the phenotype influenced by such genes. Races certainly vary in physical appearance. Nor are the differences just skin deep; there are also variations in susceptibility to disease and in the response to drugs.

The overarching similarity of all races is just what would be expected, given that the ancestral human population existed only 50,000 years ago, and given that human nature must to a great extent have been molded before the ancestral dispersal, since all its principal features are found universally. Proof of the continuing unity of the human family is that people of different races have no difficulty in interbreeding, and that the members of any one culture can, absent discrimination, function in any other.

But the existence of considerable variation between races should not be any surprise either, given that the human family has long been split into separate branches, each of which has evolved independently for up to 50,000 years or more, buffeted in different directions by the random forces of genetic drift and the selective pressures of different climates, diseases and societies.

Study of racial variation is not yet a scholarly pursuit, except in the area of medicine, and even there is not without controversy. Physicians who study racial disparities in medicine are well aware that many social attributes, such as poverty or lesser access to health care, track along with race. These factors, just as often as genetics, may be the explanation why African Americans, say, suffer a greater burden of certain diseases than white patients. But to ignore race altogether, as some argue should be done, would blind researchers to many findings of value, both social and genetic. A physician cannot tell if his black patients receive worse health care than whites unless he has first noted which race they belong to.

As for genetic contributions, BiDil would never have been discovered if Jay Cohn had not analyzed the response of African American patients to the drug. The discovery of an important drug for an underserved community might seem unalloyed good news. Nonetheless, some African Americans greeted BiDil with a distinct lack of enthusiasm because of a wider concern. These spokesmen fear that if African Americans are defined genetically, even for the benign purpose of medicine, the public may associate them with less reputable attributes, such as propensity to crime. "If you think in terms of taxonomies of race, you will make the dangerous conclusion that race will explain violence," Troy Duster, a sociologist at New York University, said in objecting to race-based medicine.[257]

Understandably enough, any suggestion of a genetic basis for racial differences can engender strong passions. Disputes have long swirled around intelligence tests, which at present show differences between the various races of the United States. There is broad overlap between all populations but in terms of average score, Asian Americans come out somewhat higher than people of European ancestry, while African Americans score lower. While this fact is generally accepted, there is little agreement as to the reason. Some psychologists claim that IQ tests measure general intelligence, which they believe is in substantial degree inherited, and that the tests predict performance in later life. Others see the tests as evidence only of differences in education and other cultural advantages, and deny that any genetic explanation is applicable. This dispute, whose merits lie beyond the scope of this book, has long made the study of race controversial.

A less vexed instance of racial differences is provided by sports records. Some 95% of the top times in sprinting are held by West Africans, or African Americans who trace their ancestry to West Africa, according to Jon Entine. Entine, a filmmaker, made a documentary about black dominance of sport and then wrote a book, *Taboo*, so called because of the obloquy rained down on anyone who suggests a genetic basis for any aspect of race. West Africans' dominance of sprinting is so complete that "all of the thirty-two finalists in the last four Olympic men's 100-meter races are of West African descent," Entine writes.[258]

In middle distances, of 5,000 to 10,000 meters, it is not West Africans but men from Kenya in East Africa who dominate. Kenyans hold the top 60 world times in the 3,000-meter steeplechase and more than half the top times in the 5,000 and 10,000 meters. Within Kenya, most of these winning runners are Kalenjin speakers of the Great Rift Valley region, particularly a small population called the Nandi. The Nandi, who comprise less than 2% of Kenya's population, have produced half of Kenya's Kalenjin-speaking athletes and 20% of all the winners of major international distance running events.[259]

Interestingly, the Kenyans have tried to extend their domination of the middle distance to the sprint, but with serious lack of success: the best Kenyan time in the 100 meter sprint ranks about 5,000th on the all-time list.[260] This suggests they possess some quality of specific relevance to middle-distance running.

Entine notes the many social factors that have helped determine dominance in sports, at least in the past. Basketball in the 1930s was dominated by Jewish players, and the sportswriters who speculated about some Jewish genetic suitability for the game were way off base. But many sports, particularly track events, are now much more open to all comers, regardless of race or social background. Despite the hard training and other factors that make a great athlete, there is likely to be some genetic component behind the spectacular dominance of West African athletes in sprinting and East Africans in middle distance events, in Entine's view. John Manners, an author of books on Kenyan runners, also favors a genetic explanation for the prowess of the Kalenjin athletes whose record, he asserts, "marks the greatest geographical concentration of achievement in the annals of sport."[261] The Kalenjin, martial Nilotic pastoralists from Ethiopia, have lived for centuries at altitudes of 2,000 meters or more and marry mostly among themselves. They have a particular custom that could have acted as a genetic selection mechanism favoring strong runners, Manners writes. It has to do with cattle-raiding, of which the Kalenjin were for a long time the leading practitioners.

While some might call that theft, the Kalenjin regarded their actions as repossession of property that was theirs by divine right but had inadvertently fallen into others' hands. The repossession procedure often required journeys of more than 100 miles so that the livestock could be far away before their ex-custodians realized their loss. "The better a young man was at raiding—in large part, a function of his speed and endurance—the more cattle he accumulated," Manners writes. "And since cattle were what a prospective husband needed to pay for a bride, the more a young man had, the more wives he could buy, and the more children he was likely to father. It is not hard to imagine that such a reproductive advantage might cause a significant shift in a group's genetic makeup over the course of a few centuries."[262] As Manners emphasizes, this is a speculation, not a proof, as to how the Kalenjin got to be so fleet of foot.

International sports events are an effective way of showing up even slight differences between races, and between ethnic groups within races, because of the way that physical characteristics tend to be distributed in a population. Most members of a population are of average height, very few are of dwarf or giant stature. If one population is very slightly taller than another, the difference might hardly be noticeable in comparing average members of each population. But if you hold a competition for the ten tallest people, all 10 may come from the slightly taller population since in this case it is the extreme, not the average, that is being compared.

The fact that different races or ethnic groups tend to excel at different sports—Africans at track, Chinese at ping pong, Europeans at weightlifting—is not proof in itself of any genetic component but just a starting point that hints at possible genes to look for.

Genes versus Geography

Even though the individual members of every race may be much the same, human societies differ considerably in their levels of technology and organization. Some societies, like those of New Guinea, are just emerging from Stone Age cultures, while others, like those of Finland or Taiwan, are highly educated and lead in manufacturing sophisticated goods for the global economy. Is the difference solely because New Guineans were dealt a bad hand in terms of geography and resources, or could there be some genetic difference, maybe in the nature of sociality, that helped keep New Guineans and others in the Stone Age while propelling other peoples on a quite different trajectory?

Jared Diamond of the University of California, Los Angeles, has advocated a geographical answer to this question. In his book *Guns, Germs, and Steel* he argues that because more domesticatable species of plant and animal existed in Eurasia, agriculture got started there first, giving Europeans a head start in economic development. Accustomed to living in crowded environments, Europeans built up immunity to many diseases, including those contracted from their domestic animals, such as influenza, measles and smallpox, and these diseases were devastating to nonurban peoples on other continents.

In Diamond's view, it was the economic head start and the germs, not any inherent difference in abilities, that enabled Europeans to conquer other peoples. "History," he says, "followed different courses for different peoples because of differences among peoples' environments, not because of biological differences among peoples themselves."[263]

As Diamond explains, having spent many years studying the birds of New Guinea, he came to know the inhabitants well and was impressed with their evident intelligence. This led him to doubt the findings of the IQ testers in America, where "numerous white American psychologists have been trying for decades to demonstrate that black Americans of African origins are innately less intelligent than white Americans of European origins."

In fact, New Guineans, in Diamond's view, are probably more intelligent than Westerners, and the reason, he says, is genetic. The chief selective pressure on Westerners was the need to acquire resistance to the disease rampant in their crowded communities, whereas in New Guinea, where the chief cause of death is war, murder or starvation, one needed one's wits to survive: "Natural selection promoting genes for intelligence has probably been far more ruthless in New Guinea than in more densely populated, politically complex societies, where natural selection for body chemistry [that is, immunity to disease] was instead more potent," Diamond explains. And hence, "in mental ability New Guineans are probably genetically superior to Westerners."

But if the New Guineans had the smarts, why was it the dumber, disease-ridden Westerners who figured out how to escape from the deadening cycle of Stone Age tribalism and perpetual warfare, a problem the New Guineans never cracked? Because Westerners lucked out in their geography, Diamond argues. Eurasia had a greater absolute number of plant and animal species and more of them proved suitable for domestication. Because species are adapted to climatic zones, domesticated crop plants and animals could be shared along lines of latitude, enabling Europeans to assemble packages of agricultural species and get a head start on the farming revolution. This advantage, slight enough 10,000 years ago, steadily accumulated to the point that by AD 1500 great civilizations had arisen in both halves of the Eurasian land mass, while much of the rest of the world had yet to clamber out of tribalism and illiteracy.

The Chinese then lost their technological edge, also for a geographical reason, in Diamond's view: the connectedness of the Chinese mainland allowed one ruler to dominate and make irreversible errors, like destroying the Chinese fleet, whereas in Europe, with its balkanization and competing statelets, diversity thrived and the best idea had a better chance of winning out. By colonial times, this left Europeans as the winners, thanks to their superior geography.

Single cause explanations generally make historians roll their eyes but the boldness and ingenuity of Diamond's thesis certainly puts geography more on the map than it was before. Yet does genetics have no role at all in shaping human history?

Many readers who like the political implications of Diamond's thesis—that Western dominance is an accident of geography and therefore no race is better than any other—may skip over his premise of New Guinean genetic superiority. But if New Guineans adapted genetically by developing the intellectual skills to survive in their particular environment, as Diamond says is the case, why should not other populations have done exactly the same?

In attributing western advance solely to geography, while tacitly excluding the genetic explanation invoked for the New Guineans, Diamond focuses on the development of agriculture. But, as noted in chapter 7, archaeologists now believe that in the Near East sedentism came long before agriculture: first people settled down, abandoning the foraging way of life. Then they took to cultivating wild plants. Then, probably by accident, they developed domestic varieties of plant and animal species. The critical step was not domestication, but sedentism. This finding would seem to undercut an important part of Diamond's case because, unlike the case with agriculture, it's harder to see any geographical reason why sedentism should have risen in one society and not another. Given that the human form was undergoing another genetically driven change around this time, the gracilization of the skull and skeleton, a genetic explanation for sedentism would not be so implausible. People such as the Natufians perhaps responded to their environment with a different kind of sociality that enabled them to abandon the foraging way of life and settle down in fixed communities.

If sedentism was indeed prompted by an evolutionary change, it was one that may have occurred independently in different populations, as has happened with properties like pygmy stature, lactose tolerance and doubtless many others.

Such genetic adaptations, if they occurred, could not spread through the world's population like wildfire, since it can take many generations for gene frequencies in a population to change. Instead, they would take place at different rates in different populations. This wide spread in start times for the forager-settler transition could help explain why human societies throughout the world have attained such different levels of development.

Emergence of Human Races

When did today's continental-based races start to emerge? Presumably people started adapting independently to different environments as soon as the ancestral population dispersed 50,000 years ago. Yet skull types throughout the world remained much the same throughout the Upper Paleolithic period, and it seems that those typical of today's races did not appear until about 12,000 to 10,000 years ago.[264] The Han Chinese originated from a small population that emerged around that time and then expanded very quickly, presumably at the expense of its neighbors. The same appears to be true of Caucasians (the peoples of Europe, India and the Near East), whose skulls resemble those of earlier Europeans, as if derived from them, but also differ from them. These earlier Europeans have larger skulls, with heavier jaws and brow ridges, and "should probably not be lumped with living Europeans in a 'Caucasoid' race," says the paleoanthropoligist Richard Klein.

It is tempting to see the origin of today's Caucasians and East Asians in the people who lived in the northern latitudes of Europe and Siberia respectively some 20,000 years ago. As mentioned earlier, these populations would have been driven southward by the advancing glaciers of the Last Glacial Maximum. Since all but the southern fringes of the Eurasian continent were converted to polar desert or tundra, the heartlands of both Europe and Asia would probably have been depopulated (see figure 6.2).

When the glaciers began their final retreat 15,000 years ago, the former northerners in both halves of the Eurasian continent would have recolonized the abandoned latitudes. In this way both Europe and East Asia would have been dominated by peoples originating from groups that 5,000 years earlier had been small populations at some northern extremity of the human population range.

A third continental race, that of American Indians, is descended from a few groups of Siberian ancestors, so also represents the expansion to continental size of a small population.

Europeans, East Asians and American Indians seem therefore to be three comparatively young races, and the two other continental races, Australasians and Africans, may be somewhat older in the genealogical sense (that is, have longer branches to the common origin). But Africans and Australian aborigines have had just as long to evolve and, aside from having retained darker skins, may be as different from the ancestral people as are the three races that emerged in northern latitudes.

Races arise from the fact that after a population splits, its two halves continue to evolve but along independent paths. These population splits leave their mark not only in the genes but also in language. Like the genes, language is in constant flux, and diverges into daughter tongues after a population goes separate ways. At the time of the ancestral population, there was a human family that spoke, perhaps, a single mother tongue. Having considered the division of the human family into races, it is now time to look at the parallel fragmentation that has occurred in language.

10

LANGUAGE

It may be worth while to illustrate this view of classification, by taking the case of languages. If we possessed a perfect pedigree of mankind, a genealogical arrangement of the races of man would afford the best classification of the various languages now spoken throughout the world; and if all extinct languages, and all intermediate and slowly changing dialects, had to be included, such an arrangement would, I think, be the only possible one. Yet it might be that some very ancient language had altered little, and had given rise to few new languages, whilst others (owing to the spreading and subsequent isolation and states of civilisation of the several races, descended from a common race) had altered much, and had given rise to many new languages and dialects. The various degrees of difference in the languages from the same stock, would have to be expressed by groups subordinate to groups; but the proper or even only possible arrangement would still be genealogical; and this would be strictly natural, as it would connect together all languages, extinct and modern, by the closest affinities, and would give the filiation and origin of each tongue.

CHARLES DARWIN, ON THE ORIGIN OF SPECIES

ACROSS THE WORLD, linguists estimate, there are some 6,000 different languages. All are descendants of older languages that are no longer spoken. In a few cases these parent languages have survived in written form, like Latin, or can be reconstructed from their descendants, like proto-Indo-European, the inferred ancestor of a vast family of languages spoken from Europe to India.

The 6,000 languages, in other words, are not an unrelated miscellany but all belong to various branches of a single family tree of human languages. Those branches must presumably have converged at their trunk to a single language, the first ever spoken, which was perhaps the mother tongue of the ancestral human population.

If so, it should be possible to draw up a genealogy of the world's languages, showing their tree of descent from the mother tongue. As Darwin perceived, such a tree should be recognizably similar to a parallel tree showing the emergence of human races from the ancestral population. And if a tree of language could be interwoven with a genetic tree of human populations, and the two trees linked to the various cultures discovered by archaeologists, a new and unified framework would be created for understanding all of human prehistory.

One immediate obstacle to this grand synthesis is that most historical linguists believe language trees cannot be constructed farther back than a mere 5,000 years from the present, or perhaps 10,000 years at most. Geneticists, however, are not so pessimistic. They have developed sophisticated statistical techniques for constructing genetic trees and believe the same approach should work for languages.

The geneticists' methods, if they work, may help resolve several long-running disputes in historical linguistics. Foremost among these is the question of the unusual distribution of the world's languages.

Language Spread Zones and Mosaic Zones

Across the United States a single dominant language is spoken. New Guinea, by contrast, has some 1,200 languages, a fifth of the world's total, jammed into an area a quarter the size of the continental United States. Why should the linguistic situations be so different?

Linguists call a large area dominated by a single language a spread zone. An area parceled into many small regions, each of which has its own language, is a mosaic zone. Most of the world's language zones fall into one or other of these two patterns, and throughout history there seem to have been occasional alternations between them. The forces that generate mosaic zones and spread zones are significant shapers of history and culture.

Mosaic zones arise in part because language mutates so rapidly, even from one generation to another, that in only a few centuries it passes beyond easy recognition. Just six hundred years later, the English of Chaucer seems half way toward a foreign language. Within a language there are dialects, that often change from village to village and were probably even more distinctive in days when people seldom traveled far from home. Even in England, up until the late 1970s, speakers could be located by their accent to an area as small as 35 miles in diameter.

This variability is extremely puzzling given that a universal, unchanging language would seem to be the most useful form of communication. That language has evolved to be parochial, not universal, is surely no accident. Security would have been far more important to early human societies than ease of communication with outsiders. Given the incessant warfare between early human groups, a highly variable language would have served to exclude outsiders and to identify strangers the moment they opened their mouths. Dialects, writes the evolutionary psychologist Robin Dunbar, are "particularly well designed to act as badges of group membership that allow everyone to identify members of their exchange group; dialects are difficult to learn well, generally have to be learned young, and change sufficiently rapidly that it is possible to identify an individual not just within a locality but also within a generation within that locality."[265]

In warfare, dialect may serve to distinguish friend from foe. When Jephthah and the men of Gilead defeated the Ephraimites, guards were posted to prevent the survivors escaping back across the Jordan. "And it was so," the bible recounts in chilling detail, "that when these Ephraimites which were escaped said, Let me go over; that the men of Gilead said unto him, Art thou an Ephraimite? If he said, Nay; Then said they unto him, Say now Shibboleth: and he said Sibboleth: for he could not frame to pronounce it right. Then they took him, and slew him at the passages of Jordan: and there fell at that time of the Ephraimites forty and two thousand."[266]

On Easter Monday in 1282, the people of Sicily rose up against the occupying French troops of Charles of Anjou. "Every stranger whose accent betrayed him was slaughtered, and several thousand Frenchmen were said to have been killed in a few hours," the historian Denis Mack Smith writes of the massacre known as the Sicilian Vespers.[267] The linguistic challenge was to say "ceci" (pronounced "chaychee"), the Italian word for chickpeas.

The mutability of language reflects the dark truth that humans evolved in a savage and dangerous world, in which the deadliest threat came from other human groups. Mosaic zones presumably come into being when small tribal groups coexist for a long time in the same place, with none being able to overrun the others. Even if the original settlers all speak the same language, dialects quickly evolve in each group's territory, as a badge of identity and a defense against outsiders. The longer this situation lasts, the greater the diversity of languages that are spoken.

New Guinea, a premier example of a mosaic zone, appears to have so many languages because it has been stable for a very long time. There seem to be two principal language families, Trans Guinea in the central mountains, and Austronesian languages spoken around the coastal plains. Trans Guinea is the language of earlier settlers, possibly even the original ones who arrived 40,000 years ago, while Austronesian is thought to have arrived with rice-growing seafarers who expanded from Taiwan throughout the islands of the Pacific.

Each of New Guinea's languages is spoken, on average, by some 3,000 people living in 10 to 20 villages. Tribal competition, as well as the deeply forested mountains and valleys, is one reason for the extreme balkanization. "Political fragmentation is a fact of life in New Guinea communities," writes William Foley, an expert on the island's languages. "Unlike most of Eurasia and much of Africa, the region does not have a history of state formation, either of empire or nation type. The basic unit of social structure is the clan, and competition between clans is the basic arena in which political life is played out."[268] Thus three factors that have shaped the island's rich mosaic of languages are competition, the inability of any one language group to dominate the others, and a long period of time for diversification to occur.

The same process may have occurred on a worldwide basis after modern humans first left the ancestral homeland. Linguistically, a single worldwide spread zone would have been created, because the small group that left Africa presumably spoke a single language. But that spread zone would have been occupied by mutually hostile tribes who deterred travel across their territory by any who didn't speak their tongue. Over the generations this worldwide spread zone would have crystallized into a mosaic zone of increasingly divergent languages. New Guinea and parts of Australia may represent the remnants of that ancient mosaic zone. Given the territoriality of early people, reinforced by language barriers, it is little wonder that the world's population has been so immobile, at least as reflected in its genetic composition, until recent times.

Discovery and exploitation of a new, uninhabited territory would open up a new language spread zone, though that too, once occupied, would gradually fragment into the mosaic pattern. South America, with its many Amerind-derived languages, is a recently created mosaic zone. But two areas of the world have been inhabited so recently that they still look like spread zones. One is Polynesia, the other is that of the arctic regions, first occupied when the Inuit peoples developed the technology for living there.

Once a spread zone has crystallized into a mosaic zone, what forces can make it revert to a spread zone? Three possibilities are climatic disaster, a transition to agriculture, and warfare.

If a large land area is wiped clean of people, those who recolonize the empty lands will create a spread zone of their own language. The Last Glacial Maximum depopulated the northern part of the Eurasian continent between 20,000 and 15,000 years ago. Those who returned could have been the speakers of the ancient language that preceded proto-Indo-European and other large language families. This postulated ancient superfamily is called Nostratic by some scholars, and proto-Eurasiatic by the linguist Joseph Greenberg. Or possibly it was the Younger Dryas cold snap, beginning around 13,000 years ago, that paved the way for Eurasiatic and its daughter languages.[269]

Another major perturber of mosaicism may have been agriculture. Colin Renfrew of the University of Cambridge and other archaeologists, such as Peter Bellwood of the Australian National University in Canberra, believe that from each center where agriculture was first developed, populations may have expanded outward, spreading their languages with them.

Bellwood and the geographer Jared Diamond argue that no fewer than 15 major language families are the result of farmers expanding from the first centers of agriculture.[270] In some cases a single center spawned several different language families, they suggest. Presumably this could have happened if an agricultural center covered several highly diversified languages in a mosaic zone, all of whose populations were amplified by the new farming technology.

Diamond and Bellwood propose that the center of agriculture in the Near East was the source of at least two major language families. One was the Indo-European family of languages. Another was Afroasiatic, which they say spread southwest into Africa. A third could have been Dravidian which, even before Indo-European, had expanded in a southeasterly direction into India. (Dravidian is distantly related to Elamite, an ancient language spoken in southwestern Iran; the eastern branch of Indo-European presumably arrived in India later, pushing the Dravidian-speakers southward.)

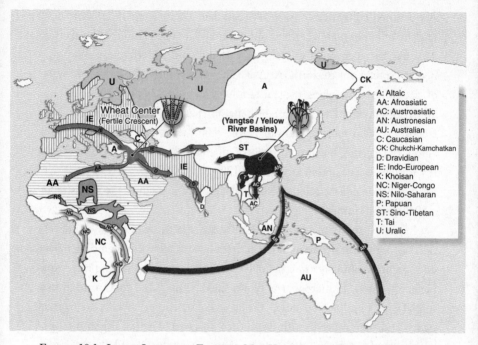

FIGURE 10.1. LARGE LANGUAGE FAMILIES MAY HAVE ARISEN THROUGH FARMING.

The language/farming hypothesis holds that populations expanded from the regions where agriculture was invented, spreading their languages with them. If several languages were spoken within such a region, all could be exported from it. The Indo-European and Afroasiatic languages may have originated in the wheat center, according to the hypothesis, and perhaps Dravidian too. The Sino-Tibetan, Tai and Austroasiatic language families are proposed to have spread from the rice center, along with Austronesian, whose speakers reached Taiwan and from there expanded across the southern oceans.

These language expansions would have taken place up to 9,000 years ago (see arrows). The map of the world, however, shows the distribution of present day language families. People speaking an Indo-European language known as Tokharian expanded into northwest China but their language is now extinct.

Also shown is the Bantu expansion in Africa, labeled for Bantu's Niger-Congo language family, which occurred some 4,000 years ago.

The proposal of the Fertile Crescent as a spawner of language families is ingenious, but the origin of each of the language families involved is a matter of dispute. In the case of Afroasiatic, linguists such as Christopher Ehret, of the University of California, Los Angeles, vigorously dispute Bellwood and Diamond's proposal that the language family originated in the Near East.[271]

A second major homeland of language families, according to the Diamond-Bellwood thesis, was the region of the Yangtze and Yellow river basins where rice was first cultivated some 9,000 years ago. The rice region, in their view, was the origin of no fewer than four different language families. Speakers of Austroasiatic, a group of 150 languages that includes Vietnamese and Cambodian, spread out to southeast Asia. They were followed by a second wave of rice farmers, speaking the Tai family of languages, which includes Thai and Laotian. Third were the Sino-Tibetan speakers. Fourth were the Austronesians, who reached Taiwan before 5,000 years ago and then set sail across the Pacific, becoming the first inhabitants of Polynesia, and finally reaching New Zealand in around AD 1200.

The Maori colonization of New Zealand was, in a sense, the final step in a 50,000 year journey.

In Africa, the Bantu language family was spread by farmers who developed an agricultural system based at first on yams and later including millet and sorghum. Starting around 4,000 years ago, in their homeland in eastern Nigeria–western Cameroon, the Bantu speakers migrated southward in two migrations. One headed down the west coast, the other crossed to east Africa and then moved south down the east coast. The latter group of migrants mingled with Nilo-Saharan speakers around the Great Lakes region of east Africa, and displaced the Khoisan speakers. Bantu languages, though just one branch of the Niger-Congo superfamily, are now spoken across a broad zone of subequatorial Africa.

Diamond and Bellwood list the Bantu expansion as being the least controversial of their 15 asserted cases of language/farming spread. But a major factor in the Bantu speakers' success, besides their farming practices, was their mastery of ironworking. Iron weapons were part of the package that made their advance through the length and breadth of Africa so irresistible, raising the possibility that warfare was also an agent of the Bantu expansion.

Warfare is a third major perturber of mosaic zones, whether by itself or combined with new agricultural techniques. During the first millennium BC, Nilotic-speaking peoples expanded southward from Ethiopia to the Great Lakes region of eastern Africa, overcoming Cushitic-speaking farmers in the Kenyan highlands. They were able to displace agricultural societies, Christopher Ehret believes, because of a superior military tradition based on assigning young men at adolescence to age sets, which served as military companies on a permanent war footing. "Over the long term of their history, most Nilotes had an institution and apparently an attitude toward war that recurrently gave them the advantage over all their neighbors, except for other Nilotic peoples, whenever conflict arose," Ehret writes.[272] (These southern Nilotes included the Kalenjin of Kenya, now renowned for their more peaceful achievement of dominating world middle-distance running records.)

The Coming of the Indo-Europeans

The Indo-European languages provide a leading test case for whether warfare or agriculture has been the dominant generator of new spread zones. The spread zone of Indo-European stretches from western Europe to the Indian subcontinent. The family includes extinct languages such as Latin, ancient Greek, Hittite and Tokharian, once spoken in northwestern China. The living descendants of proto-Indo-European include, besides English, the other Germanic languages (German, Dutch, Icelandic, Norwegian), the Slavic languages (Russian, Serbo-Croat, Czechoslovak, Polish), the Baltic languages (Latvian, Lithuanian), the Italic languages (Italian, French, Spanish, Portuguese) and the Celtic languages (Breton, Welsh, Irish).

Where was the homeland of the speakers of proto-Indo-European? When did they live? How did they and their language spread? On these questions there exist two main schools of thought, one of which asserts that Indo-European spread by the sword, the other by the plough.

In a series of papers written between 1956 and 1979, the archaeologist Marija Gimbutas identified the Indo-Europeans with the people who built the characteristic burial mounds, called *kurgan* in Russian, in the steppe area to the north of the Black Sea and the Caspian. The Kurgan people, benefiting from the domestication of the horse, started expanding from their homeland sometime after 4000 BC. By 2500 BC, in Gimbutas's estimation, these warrior-pastoralists had reached the extremities of Britain and Scandinavia, and their language developed into its many descendant tongues that are spoken from Europe to India today.

This view is supported on linguistic grounds by Ehret, who argues that if the Indo-Europeans had been peaceful farmers, many words to do with cereals should trace back to them. But Indo-European literatures are full of allusions to fighting. "We find preserved in early myths and legends almost everywhere among Indo-Europeans a glorification of battle, and particularly of death in battle, not entirely unknown elsewhere in the world, but of an intensity not often matched. We also find widely in these stories a division of society that singles out warriors as an elite group," Ehret says.

A rival hypothesis was proposed in 1987 by the archaeologist Colin Renfrew.[273] He argued that the Indo-Europeans must have been the first farmers, and that they spread out from their homeland because the new agricultural techniques allowed the population to grow and therefore expand. Looking to the archaeological evidence bearing on the spread of agriculture, Renfrew placed the homeland of the first Indo-European speakers in Anatolia, now Turkey, the region where some of the earliest Neolithic settlements have been found. Because the Neolithic revolution started expanding through Europe around 9,500 years ago, Renfrew's hypothesis required the Indo-European languages to have arrived several thousand years earlier than implied by Gimbutas's Kurgan warrior theory and indeed than the date favored by most historical linguists.

It seemed for a time that genetics might decide the issue. The first genetic insight into the peopling of Europe came from Luca Cavalli-Sforza of Stanford University. Working just with the protein products of genes, since DNA sequencing was not then available, he showed there was a genetic gradient, based on 95 genetic markers, that spread across Europe in a southeast to northwest direction. He and the archaeologist Albert Ammerman suggested the gradient was caused by Neolithic farmers moving across Europe in a slow wave of advance. Although the farmers were assumed to intermarry with the existing foragers, giving rise to the observed genetic gradient, the basic engine behind the wave of advance was assumed to be the population growth of the more numerous farmers.[274]

This idea lent serious but not conclusive weight to Renfrew's theory. Cavalli-Sforza noted that several other genetic gradients emerged from his data besides the one possibly associated with farmers from the Near East. Another gradient suggested a flow of genes westward from the steppe area above the Black Sea. This gradient "supports Gimbutas' hypothesis," he and his coauthors said, just as the first gradient supported Renfrew's.[275]

New assessments of population numbers have undercut Renfrew's original idea that population growth was the engine of Indo-European expansion. The archaeologist Marek Zvelebil, of the University of Sheffield in

England, writes that "Demographically, there is no evidence for population pressure sufficient to encourage first farmers to migrate, nor is there evidence for rapid population growth. Archaeological evidence does not record rapid saturation of areas colonized by Neolithic farmers, or demographic expansion [with one possible exception]."[276]

But Renfrew's theory could still be correct even if Indo-European-speaking farmers did not overwhelm the indigenous population of Europe. The farmers' language could have been adopted by the European hunter-gatherers along with the new agricultural technology. In terms of population numbers, relatively few farmers entering Europe from the Near East could have had a catalytic effect in spreading both their language and their farming techniques. Perhaps they bought or captured extra wives from the Paleolithic inhabitants, and the next generation moved a few miles farther into Europe, also adding wives from the existing forager population. The farther this wave of farmers advanced into Europe, the more its Neolithic genes would get diluted with Paleolithic genes. But regardless of the shifting composition of the genetic pool, each generation of farmers would speak the language of its parents' community, presumably Indo-European.

In this way, the new farming techniques would have triggered a language change throughout the area to which they were applied, but with only a small number of Anatolian immigrants relative to the indigenous forager population. This could explain how it is that Europeans speak Indo-European languages yet carry only 20% or less of the genes of those assumed to have introduced the languages.

Can Languages Be Dated?

European genetics seems at present compatible with both theories of Indo-European spread. A more decisive test would be to put a date on when proto-Indo-European was spoken, since the two theories imply very different times of expansion. The Kurgan warrior expansion started some 6,000 years ago, the spread of farming from the Near East some 9,500 years ago.

The dating of languages is not yet a settled science. One approach is to estimate the rate of historical change in a group of languages by analyzing similarities in vocabulary. Glottochronology, one version of this method, depends on estimating the percentage of cognates that two languages have in common. (Cognates are words derived from a common ancestor; *apple* is a cognate of German's *Apfel* but not of French *pomme*.)

The cognates that glottochronologists examine are not chosen randomly but belong to special vocabularies, drawn up by the method's inventor, Morris Swadesh, from items that are particularly resistant to linguistic change. These include words for numbers, pronouns and parts of the body. A Swadesh list of 100 words is the most commonly used.

In comparing two languages, a linguist will decide how many Swadesh-list words in each are true cognates with each other. The fewer cognates, the longer ago the languages diverged, and there are various methods of translating the percentage of matching cognates into a date of language split. In Ehret's view, a 5% match indicates a language split of about 10,000 years ago, a 22% agreement means a divergence around 5,000 years ago, and two languages that parted ways only 500 years ago will retain 86% of their Swadesh-list vocabulary in common.

Given the simplicity of the method, glottochronology can produce surprisingly plausible dates. But it has flaws. Linguists have put considerable effort into criticizing glottochronology, perhaps more than in trying to get it to work better. The result has been continuing disagreement among linguists as to whether it is a usable technique. At a conference held at Cambridge University in 1999, opinion ranged from one extreme to the other. Robert Blust, of the University of Hawaii, gave a paper explaining why the glottochronology kind of method "doesn't work" for Austronesian languages, and James Matisoff, of the University of California, Berkeley, talked about "the uselessness of glottochronology for the subgrouping of Tibeto-Burman." They were followed by Ehret, who explained how well glottochronology works for dating language splits in the Afroasiatic family.[277]

Historical linguists are much more enthusiastic about a quite different dating technique called linguistic paleontology. The idea is to reconstruct words for objects of material culture in a language family and date the language by noting the times at which such objects first appear in the archaeological record.

In many Indo-European languages, for example, there are words for wheel that are clear cognates of each other. Greek has *kuklos* (a word that is also the origin of circle), Sanskrit *cacras*, Tokharian *kukäl*, and Old English *hweowol* (initial "k"s in proto-Indo-European turn to "h" sounds in the Germanic family branch). Since the daughter languages of proto-Indo-European have cognate words for wheel, they must be derived from a common source, and linguists assert that this was the proto-Indo-European word for wheel, which they reconstruct as *k^wek^wlos* (the asterisk indicates a reconstructed word).

Now, the earliest known wheels in the archaeological record date from 3400 BC (5,400 years ago). The proto-Indo-European language must have split into its daughter languages sometime after this date, the argument goes, since how else could the daughter languages, spoken over an enormous region, all have cognate words for wheel?

Similar arguments can be made for words like yoke, axle, and wool. Work on this issue by linguists like Bill Darden of the University of Chicago has encouraged many linguists in their belief that Indo-European was a single language as recently as 5,500 years ago and that its daughter languages could not have come into existence until after this date.[278]

Linguistic paleontology is an ingenious exercise of the linguist's craft. But it has two conceptual weaknesses. One is that a splendid new invention like the wheel is likely to spread like wildfire from one culture to the next, carrying its own name with it. Linguistic paleontologists claim they can spot such borrowed words. It's true that "Coca-Cola" is easy enough to recognize as a foreign borrowing in many languages, but the more ancient the borrowing, the more a word may take on the coloration of its host language. One of the criticisms linguists level at glottochronology is that it is confounded by unrecognized borrowed words.

Another weakness in linguistic paleontology is the danger of constructing highly plausible words that didn't, in fact, exist. Related words for bishop exist in Greek (*episkopos*), Latin (*episcopus*), Old English (*bisceop*), Spanish (*obispo*) and French (*evêque*), from which the proto-Indo-European word **apispek* for bishop could be reconstructed; but of course, in a language spoken at least 5,000 years ago, no such word existed. As for wheel, proto-Indo-European is thought to have had a word *k^wel*, meaning to turn or twist, of which *k^wek^wlos* is assumed to be a duplication. But it could be that proto-Indo-European had no word for wheel, and what happened was that its daughter languages each independently used their inherited *k^wel*/turn words to form their own words for wheel. In which case proto-Indo-European could have been spoken thousands of years before the invention of the wheel.

A New Date for Proto-Indo-European

A better, more systematic way of dating languages has long been needed, and biologists hope they may have provided it by adapting one of their own methods for drawing phylogenetic trees. The favored approach is called a maximum likelihood method because it asks what is the most probable shape of tree to account for the observed data. In the case of language families, the data are each language's list of Swadesh words, along with a designation of which are cognates and which are not.

The idea of applying a maximum likelihood method to language history was laid out by Mark Pagel, an evolutionary biologist at the University of Reading in England. Pagel showed that with a list of just 18 words he could generate a maximum likelihood tree for 7 languages (Welsh, Romanian, Spanish, French, German, Dutch and English) that was the same as the tree constructed by linguists with purely linguistic techniques.[279]

The method has now been further developed by Russell D. Gray, an evolutionary biologist at the University of Auckland in New Zealand. Gray has carefully analyzed the problems of glottochronology and adapted the method so as to address them. One of the problems is unrecognized borrowing. Unrecognized loan words make languages appear younger than they are. But they also knit the side branches of a language together, making a netlike structure. Netlike structures can be tested for and the offending words eliminated.

Another problem that has vexed glottochronology is that languages may evolve at different rates. Both modern Icelandic and Norwegian are known to have evolved from Old Norse, which was spoken between AD 800 and 1050. Norwegian and Old Norse have 81% of their Swadesh list words as cognates, correctly implying a separation of 1,000 years ago. But modern Icelandic, which has been much more isolated, shares 99% of its words with Old Norse, wrongly implying the two languages separated only 200 years ago.[280] Rate variation can be taken account of in the maximum likelihood approach, essentially by choosing trees with the minimum amount of variation necessary to fit known dates of language divergence.

The mathematical techniques for addressing both word borrowing and variation in evolution rate were available because biologists had encountered the same two problems in drawing up trees based on DNA data. As with languages, some genes evolve at faster rates than others. And just as words may be borrowed instead of inherited, an organism may acquire genes through borrowing as well as by inheritance; bacteria, for instance, transfer packets of genes to each other, which is why they so quickly acquire genes for resistance to antibiotics.

In one maximum likelihood approach currently favored by biologists, called the Bayesian Markov chain Monte Carlo method, the DNA sequences of various genes are fed into a computer that generates a large number of possible trees by which the genes might be related. The program samples the classes of tree that seem most promising (there are far too many for even the fastest computer to examine each one), and then repeats the whole process a large number of times. At each iteration there are fewer promising trees, and eventually the process will converge on a single, most probable tree to account for the data.

With this powerful tree-drawing technique, Gray and his colleague Quentin Atkinson have constructed a family tree of Indo-European. For data, he relied on a 200 word Swadesh list for 84 Indo-European languages drawn up by the linguist Isidore Dyen, to which he added data from three extinct languages (Hittite and the two versions of Tokharian, known as Tokharian A and B).

Gene trees can often be anchored in real time by matching a date from the fossil record to one of the tree's branch points. The same can be done with maximum likelihood trees constructed for languages. Having found the statistically most likely tree to account for the Indo-European data, Gray then constrained certain branch points in the tree to fit attested historical dates for divergence of certain languages. Hittite must have been a separate language by 1800 BC, the date of the oldest known inscription. Greek must have been separate by 1500 BC, the date of the Linear B inscriptions. Latin and Romanian started to diverge when Roman troops withdrew south of the Danube in AD 270.

Altogether Gray plugged in 14 known dates, constraining the tree to fit itself to the dates in the most statistically probable way. Because the branch lengths of the tree are proportional to elapsed time, anchoring the tree to

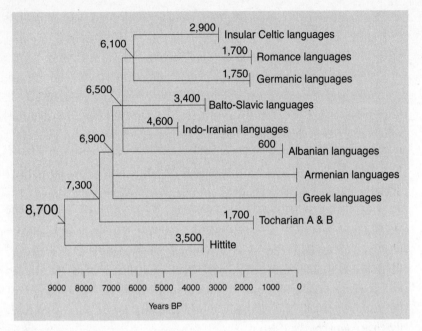

FIGURE 10.2. A GENETICIST'S TREE OF THE INDO-EUROPEAN LANGUAGE FAMILY.

A tree of Indo-European was constructed by Russell Gray and Quentin Atkinson using an advanced statistical method. Because the tree is anchored to 14 known dates of recent language origin, the dates of its ancient branch points can be estimated. Figures show the years before the present at which languages split apart.

According to the Gray-Atkinson tree, the original language, called proto-Indo-European by linguists, split 8,700 years ago into the two branches, of which the first led to Hittite and the second to all the other Indo-European languages. The early date assigned to proto-Indo-European suggests that it was the language of the people who introduced farming into Europe from the Middle East.

English is a member of the Germanic group of languages, as are Dutch, Swedish and Icelandic. The Romance language family includes French, Italian and Spanish. Russian, Czech and Lithuanian are among the members of Balto-Slavic. Hittite, now extinct, was the language of the Hittite empire in what is now Turkey; Tokharian was spoken in western China.

historical events allows all the other branch points in the tree to be dated. Gray's tree was published in *Nature* in November 2003, with a terse description of the rather complex methodology behind its construction.[281] The first reaction of many historical linguists was that he had done nothing new because his tree of Indo-European was just like theirs. But that very fact, in Gray's view, was the best possible validation of his method.

The novel feature of his tree was not its shape but its dates. They were very different from anything the linguists had imagined. The tree showed that proto-Indo-European was spoken before 8,700 years ago, the date at which it underwent its first split, when the branch leading to Hittite split off from all the rest. This date is nearly 3,000 years older than the 5,500 to 6,000 years ago date favored by many historical linguists for the breakup of Indo-European.

Gray's dates, if correct, are somewhat revolutionary because they show the roots of Indo-European are far older than expected and that language can be traced back far deeper in time than most linguists think likely. Moreover, a reliable dating method would at last allow language change to be correlated with the information emerging from archaeology and population genetics.

Many linguists say Gray's dates can't be right, essentially because they conflict with the dates given by linguistic paleontology. But linguistic paleontology is a fuzzy technique, dependent on judgment and vulnerable to undetected borrowing and fallacious reconstructions. Gray's technique applies a sophisticated statistical method, of proven value in phylogeny, to a reliable data set, the Dyen list, which represents the fruit of Indo-European linguistic scholarship. As a pioneering approach, it may well need refinement, or turn out to have some unexpected flaw. But as compared with linguistic paleontology, it doesn't seem so obviously less credible.

Gray says he has great respect for the scholarship and methods of historical linguistics and hopes linguists will come around to taking his tree seriously, once they understand that his technique avoids the much discussed errors of glottochronology.

Using a simpler phylogenetic technique, Peter Forster, an archaeologist at the University of Cambridge, has drawn up a family tree of several Celtic languages including Gaulish, the version spoken in ancient France before the Roman conquest, as well as Welsh, Breton and Gaelic. Celtic is a major branch of Indo-European. Forster's tree implies that Indo-European had diverged around 10,000 years ago, and that Celtic had split into Gaulish and its British branches by 5,200 years ago.[282] These dates have wide margins of error, but are in the same range as Gray's.

Gray's date of 8,700 years ago for the first split in the Indo-European language tree lends considerable weight to the Renfrew hypothesis that the invention of agriculture drove the spread of Indo-European.

The implications reach beyond the specific case of Indo-European. Success of the biologists' tree building methods would mean that languages can be reconstructed back to 9,000 years ago, considerably farther back in time than many linguists have supposed. The prospects for reconstructing even older trees of human languages may not be entirely hopeless.

The Greenberg Synthesis

Gray's tree building bears on a dispute that has long divided historical linguists. The issue is how best to assess the relationships between today's languages, given that language changes so fast. The world's 6,000 living languages lie at the tips of a long-vanished tree. Can that tree be reconstructed for other families besides Indo-European? Can these families be grouped into superfamilies so as to reach time depths even deeper than that of Indo-European?

The classification of languages is a matter of considerable disagreement. Many linguists, being familiar with the extreme mutability of language, are skeptical of attempts to find ancient relationships between living tongues. Languages change so fast, they believe, that the number of words two diverging languages may share because of a true cognate relationship quickly dwindles to near zero. Indeed, the number of cognates may fall to the same level as the number of word resemblances that arise purely by chance. Unless that point is recognized, the incautious researcher may assert relationships where none exist. The only acceptable way of avoiding such traps, many linguists believe, is with an approach called the comparative method. The comparative method is highly reliable. Its drawback is that it is so rigorous that it does not reach very far back in time.

Other linguists believe that the comparative method is useful for confirming a postulated relationship between two languages, but is too strict to help detect such relationships in the first place. A leading figure in this school of thought has been the late Joseph H. Greenberg of Stanford University. During his lifetime Greenberg classified almost all of the world's languages, showing how they could be grouped into some 14 superfamilies.

The superfamily classifications achieved by Greenberg and his colleague Merritt Ruhlen have been greeted warmly by geneticists because these groupings of languages largely mesh with the population splits inferred from gene-based genealogies. But many linguists repudiate Greenberg's language

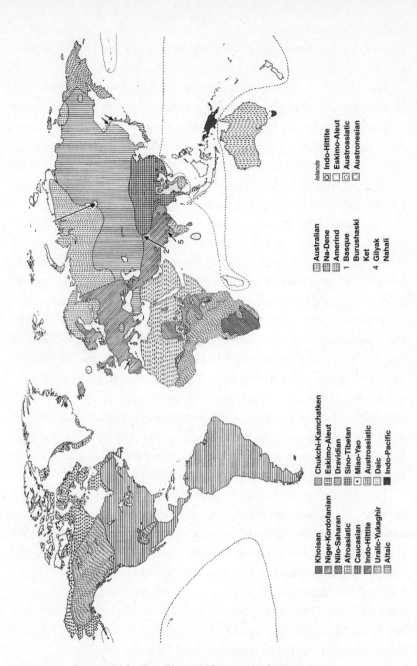

Islands
- Indo-Hittite
- Eskimo-Aleut
- Austroasiatic
- Austronesian

- Australian
- Na-Dene
- Amerind
- 1 Basque
- Burushaski
- Ket
- 4 Gilyak
- Nahali

- Khoisan
- Niger-Kordofanian
- Nilo-Saharan
- Afroasiatic
- Caucasian
- Indo-Hittite
- Uralic-Yukaghir
- Altaic

- Chukchi-Kamchatken
- Eskimo-Aleut
- Dravidian
- Sino-Tibetan
- Miao-Yao
- Austroasiatic
- Daic
- Indo-Pacific

FIGURE 10.3. THE WORLD'S LANGUAGE SUPERFAMILIES.

The language superfamilies of the Old World, as defined by Joseph Greenberg and Merritt Ruhlen. The Basque and Burushaski languages, shown by arrows 1 and 2, are entirely unrelated to their neighbors and may be relicts of more ancient languages. Ket (arrow 3) may be the mother tongue of the Na-Dene languages of North America.

families, arguing that his method is unreliable and that his work contains errors.

Greenberg was not formally an outsider to the linguistic establishment. He served as president of the Linguistic Society of America and was one of the few linguists to have been elected to the National Academy of Sciences. Aside from his work on classification, he founded a subfield of linguistics known as typology, to do with universal patterns of order in the grammatical elements of language. His 1962 article on typology is said to be the most widely cited in the history of linguistics.

Greenberg's training, however, was not in linguistics but social anthropology. He did fieldwork studying the ethnography of pagan cults among the Hausa-speaking people of west Africa, spent the years from 1940 to 1945 in the Army Signal Intelligence Corps, mostly decrypting Italian code, and after the war turned his attention to the interrelationship of African languages.

These had been largely the purview of English and French linguists who had classified them with the help of various criteria, like the physical type of the speakers, that Greenberg deemed irrelevant to language origins. He developed his own, purely linguistic method, which he later called mass comparison. It was based on comparing grammar and some 300 items of vocabulary, such as pronouns and words for parts of the body, that as Swadesh had found are less prone to linguistic change. Greenberg would fill notebooks with lists of languages down the left column and word meanings along the top, and simply search in his mind's eye for relationships.

He started out with Hausa, trying to see what other languages it might be related to by comparing common words and deciding if the languages fell into groups. Over the space of 5 years, Greenberg kept arranging the 1,500 then known languages of Africa into larger and larger assemblies, until he had grouped them into just 16 superfamilies, and finally only four. He put the odd and ancient click languages of southern Africa into the group named Khoisan. The languages of central Africa, including the widely spoken Bantu languages, he assigned to a group he called Niger-Kordofanian. He decided that the Bantu languages must have originated in west Africa, because that is where their diversity is greatest. From that it followed that the present-day Bantu languages, which are distributed down the west and east coasts of Africa, must have arisen from a migration out of the homeland that had split into two streams, one going directly down the west coast, the other crossing the breadth of Africa and then turning south down the east coast. This inference was later confirmed by archaeologists.

Greenberg's third group was Nilo-Saharan, a family of languages spoken by Nilotic peoples like the Nuer and the Dinka as well as by people of the Saharan region and by the Songhay of west Africa. The fourth group of languages, spoken in a swath across northern Africa, he named Afroasiatic. This family includes Berber of northwestern Africa, ancient Egyptian, and Semitic, a branch to which belong Arabic, Hebrew and Akkadian, the extinct language of the Assyrians and Babylonians.

Greenberg's sweeping classification of African languages has stood the test of time and is broadly accepted, although scholars continue to rearrange the furniture. The African languages are of particular interest because of their diversity and presumed antiquity. At the latest count some 2,035 are now known, of which 35 belong to the Khoisan family, 1,436 to Niger-Congo (a new name for Greenberg's Niger-Kordofanian), 196 to Nilo-Saharan and 371 to Afroasiatic.[283]

Ehret has attempted to date the period when the proto-languages of Greenberg's four groups were spoken. On archaeological evidence, he estimates that proto-Khoisan was first spoken about 20,000 years ago. The ancestral tongue of the Niger-Congo family may date back to 15,000 years ago, since a junior branch of the family had spread across the yam growing regions of west Africa from 8,000 years ago. Proto-Nilo-Saharan, on the basis of glottochronology, may be 12,000 years old.[284]

Afroasiatic is a language family of general interest since its West Semitic branch includes Hebrew, Aramaic and Arabic, the founding languages of three popular religions. Many people have assumed the ancestral homeland of proto-Afroasiatic was in the Near East, some for a miscellany of unscientific reasons, others because the Near East is a known center of early agriculture from which growing populations might have expanded into Africa, carrying their language with them. But an African origin seems more likely, in Ehret's view. Of the six major branches of Afroasiatic, five lie in Africa— Berber in northwest Africa, Chadic around Lake Chad at the southern edge of the central Sahara, Cushitic in the Horn of Africa, Omotic in the Ethiopian highlands, and ancient Egyptian.

Following the rule that the region of greatest diversity is usually the homeland, this distribution points strongly to an ancestral homeland for

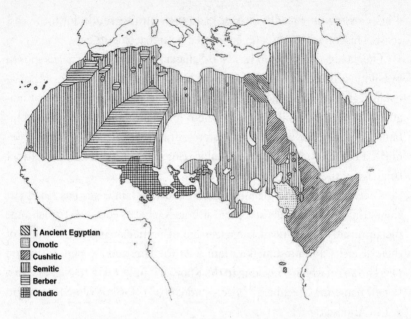

FIGURE 10.4. THE AFROASIATIC LANGUAGE FAMILY.

The major branches of the Afroasiatic language family. Arabic is now spoken in the area shown as belonging to Ancient Egyptian.

Afroasiatic somewhere in northern Africa, which the Semitic speakers left to invade the Near East, perhaps some 9,000 years ago.[285] (Later, about 7,000 years ago, some crossed back from Yemen into Ethiopia, giving it the country's principal language of Amharic.) Also pointing to an African homeland, the earliest branching of proto-Afroasiatic was into Omotic and the rest, and the second branching was into Cushitic and the rest. Since Omotic and Cushitic are both restricted to Africa, that has "put it beyond doubt that the ancestral language, proto-Afroasiatic, was spoken in Africa," writes Ehret.[286]

Though Greenberg's classification of African languages is now broadly accepted, it was for many years bitterly resisted by British Africanists. In linguistics as in other academic fields, specialists tend to resent the generalist who shows how their little patch relates to a larger order. Paul Newman, a linguist at Indiana University, recalls visiting the London School of Oriental and African Studies around 1970, some 15 years after the first publication of Greenberg's African work. He was told that it was quite safe for him to go into the common room, as long as he did not mention Greenberg's name.[287]

After his African classification, Greenberg turned his attention to the question of American Indian languages. Taking note of the archaeological findings that the Americas had been settled only recently, Greenberg expected to find far fewer language families than in Africa. But American linguists, then undergoing a splittist phase, had agreed at a conference in 1976 that no fewer than 63 independent language families were spoken in the Americas. Greenberg, using the same mass comparison method he had developed for Africa, announced there were just three—Amerind, Na-Dene and Eskimo-Aleut.[288]

Greenberg's conclusions induced the same agitation among American linguists as his African classification had among the British. And even though American linguists had generally accepted his grouping of African languages, they now assailed him with a fury that startled the population geneticists who were beginning to take an interest in his work. Luca Cavalli-Sforza, an eminent geneticist at Stanford University, wrote of his dismay at the linguists' diatribes against Greenberg.[289]

Cavalli-Sforza's confidence in Greenberg's approach stemmed from the fact that, at least in general outline, he had confirmed it by an independent approach. Before methods of DNA analysis became available, Cavalli-Sforza and colleagues had worked out a genetic family tree of the world's populations in terms of protein differences. Comparing this tree to Greenberg's list of major language families, Cavalli-Sforza showed that peoples who were grouped together on his world population tree tended to fall into the same language family, as defined by Greenberg.[290] Further analysis proved that the correspondence between the world's human population tree and Greenberg's language families was statistically significant.[291]

The Comparative Method
versus Mass Comparison

Despite Cavalli-Sforza's support for Greenberg's findings, linguists continued to assail Greenberg's work on grounds of factual errors and methodology. As even Greenberg's supporters concede, he was interested in the big picture, not the details. Numerous small errors, of the type scholars usually do their best to avoid, crept into his work. Some were errors of transcription, some perhaps the result of working in haste as he reviewed the grammar and vocabulary of hundreds of languages, transcribing everything with his own hand and usually without a graduate student to check things. Were the errors fatal, as his Americanist critics contended, or trivial, as his supporters averred? The verdict of the Africanists, who came to agree with him, is that the errors were not significant. "There are . . . more errors in data-entering than one expects in such a work," writes Lionel Bender, an Africanist at Southern Illinois University, about Greenberg's book on African languages. "Nevertheless, he got it right for the most part and his African classification culminating in the 1963 book is a tremendous advance."[292]

The larger point of Greenberg's critics was that in establishing relationships among languages he had failed to use what is known as the comparative method, the orthodox approach to classifying languages. The method is based on identifying sets of related words that change in predictable ways between members of a language family. The French and Italian words for "goat" are not particularly similar, but when compared with other words it is clear that a "k" sound in Italian corresponds with a "ch" sound in French, and a "p" in Italian corresponds with an "f" or "v" in French.[293] These sound correspondences exist because many French and Italian words are cognates, or descendants of the same parent word in their common ancestor tongue of Latin.

LATIN	ITALIAN	FRENCH	ENGLISH
capra	*capra*	*chèvre*	*goat*
caput	*capo*	*chef*	*head*
canis	*cane*	*chien*	*dog*

Once the rules of sound correspondence between contemporary languages have been established, the word in the parent language can be reconstructed. Scholars have reconstructed an extensive vocabulary in proto-Indo-European, the hypothesized ancestral tongue of many European and Indian languages. Any claim that a language is part of the Indo-European family can then be tested by seeing if its grammar and vocabulary can be derived, by the established rules, from proto-Indo-European. From the instances above, English might not seem so promising a candidate, but the initial "k" sound in Latin is known to correspond with an "h" in the Germanic group of languages, making *head* and the German word *haupt* (now a figurative word for head) cognates with Latin's *caput*. By the same rule Latin's *canis* is cognate with German's *hund* and the English word *hound*, all being derived from proto-Indo-European **kwon*.

Rigorous application of the comparative method has freed linguistics from many false etymologies and crank theories. Many linguists insist that the comparative method is the only acceptable way of testing whether languages are related to each other. This position is based on the belief that, since words change so fast, two daughter languages will soon have only a small percentage of their vocabulary in common and at this point the number of true cognates may be exceeded by chance resemblances and words that sound alike because the two languages under comparison each borrowed them from a third.

Because the signal of the true cognates is soon overwhelmed by the noise of specious ones, the roots of a family of languages, linguists say, can be traced no farther back than about 6,000 years or so, the period when most linguists believe proto-Indo-European was spoken.

Greenberg, in his method of mass comparison, did not look for sound correspondences, nor did he try to reconstruct proto-languages to confirm his findings. Hence, in the view of many linguists, his method and findings cannot be trusted.

Whatever the theoretical objections to Greenberg's method, the bottom line is the empirical question of whether or not it works. Africanists have decided it did indeed work for African languages. But this apparently persuasive circumstance has not changed linguists' views about the validity of Greenberg's method. In a recent essay on Greenberg's Afroasiatic family, Richard Hayward, of the London School of Oriental and African Studies, writes that the "only admissible evidence" for establishing that languages have a common ancestry is by the comparative method and sound correspondences. "Now it was on the basis of 'mass comparison,' rather than the comparative method, that the canon of the Afroasiatic languages was established by Greenberg, and although this methodology . . . has, in the present writer's view, come up with the right conclusions, a methodology that does not invoke the rigour of the principle stated in the last paragraph [i.e., that of the comparative method] cannot make predictions, and so falls short of true theoretical status," Hayward writes.[294] In other words, even if Greenberg got the right answer, it was by the wrong method.

If the faculty of human language were extremely ancient, and if human populations were highly mixed, the likelihood of languages on the same continent being related to each other might be small, and it would be appropriate to assume languages were unrelated unless proven otherwise. But since fully modern language probably evolved only 50,000 years ago, and since today's populations still strongly reflect the original patterns of human migration, the reverse is the case: all languages are probably offshoots of a single mother tongue and related to each other at one level or another. In circumstances where history and archaeology make language relationships very likely, such as in the Americas, a lesser standard of proof would perhaps be appropriate. It is surely in Africa, where languages have had longest to diversify, that Greenberg's mass comparison method stood least chance of success, yet it is there that linguists judge it to be most successful.

Linguists' insistence on comparative method as the only acceptable classification tool is a matter of some frustration to researchers who would like to integrate the findings of population genetics and archaeology with a linguistic tree. Without a guide from conventional linguists to deep language relationships, population geneticists tend to rely on the work of Greenberg and Merritt Ruhlen, his Stanford University colleague, as the best available guide to the overall structure of the world's languages.

The Eurasiatic Superfamily

As Greenberg worked on classifying the languages of the Americas, he realized that they must be related in some way to languages on the Eurasian continent, if indeed the Americas had been inhabited by people migrating from Siberia. So to help with the American classification, he started making lists of words in languages of the Eurasian land mass, particularly personal pronouns and interrogative pronouns.

"I began to see when I lined these up that there is a whole group of languages through northern Asia," he said in an interview in 1999. "I must have noticed this 20 years ago. But I realized what scorn the idea would provoke and put off detailed study of it until I had finished the American languages book."[295]

This was the beginning of Greenberg's next major classification, a link-

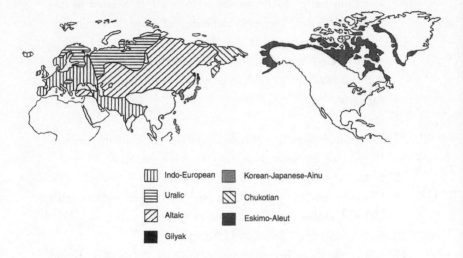

FIGURE 10.5. THE DISTRIBUTION OF EURASIATIC.

The family of Indo-European languages, according to the linguist Joseph Greenberg, belongs to a more ancient superfamily called Eurasiatic. Other members include the Uralic and Altaic families, the Korean-Japanese-Ainu group and the Eskimo-Aleut languages of North America.

ing of many of the major language families of Europe and northern Asia into a single superfamily that he called Eurasiatic. This ancestral tongue, in his view, gave rise to eight families of languages, now spoken in a great swath across northern Eurasia, from Portugal to Japan, and, since Eskimo languages too are included, from Alaska to Greenland.

The best-known member of the Eurasiatic superfamily is the language family known as *Indo-European*, which itself has 11 branches:

1. The Anatolian group, not well known because all its member languages are now extinct. Its principal member is Hittite, the language of the Hittite empire that was centered in Anatolia (now Turkey), and reached its height between 1680 and 1200 BC.

2. Armenian

3. Tokharian, a pair of languages known as Tokharian A and Tokharian B and spoken in northwest China in the second half of the first millennium AD. Though at the east of the Indo-

European range, Tokharian seems more closely related to languages of the west; the origin and history of its speakers is unclear.

4. Indo-Iranian, which includes the ancient Sanskrit as well as many modern Indian languages such as Urdu and Hindi, along with the ancient and modern languages of the Iranian region.
5. Albanian
6. Greek
7. Italic, which includes Latin and its modern descendants, such as Italian, French, Spanish, Portuguese and Romanian.
8. Celtic, which includes Irish and Scottish Gaelic.
9. Germanic, including Danish, Swedish, Norwegian and Icelandic; German, Dutch and Yiddish; and English.
10. Baltic, including Latvian and Lithuanian.
11. Slavic, the branch comprised of Russian, Polish, Czech and Serbo-Croatian.

The second major family of Eurasiatic is *Uralic-Yukaghir*, a far flung family that includes Hungarian, Finnish and Estonian in the west and many Siberian languages in the east. This family, in Greenberg's view, includes Ket, a hard to classify Siberian language that may be the source of Na-Dene, the second of the three language families of the Americas along with Amerind and Eskimo-Aleut.

Third is *Altaic*, which includes the Turkish and Mongolian language groups.

Fourth is *Korean-Japanese-Ainu*, a grouping that has no generic name; Ainu is the language spoken by the original inhabitants of northern Japan.

Fifth is *Gilyak*, the language of a dwindling number of people who live in northern Sakhalin, the large island north of Japan, and in a small region opposite Sakhalin on the Siberian mainland.

Sixth is *Chukotian*, a language family of eastern Siberia that includes Chukchi and Koryak.

Seventh is *Eskimo-Aleut*, a family spoken from Siberia to Greenland.

Eighth is *Etruscan*, an extinct language of the Romans' adversaries in ancient Italy.

Greenberg's book on the grammar of his proposed Eurasiatic family was published in 2000; the second volume, on shared vocabulary, appeared posthumously in 2002. His grouping was developed independently of Nostratic, the superfamily advocated by a Russian school of linguists, but overlaps with it to a great extent. Nostratic differs from Euroasiatic in that it includes Afroasiatic, at least in early versions, and some Nostraticists exclude Japanese and Ainu. An important difference of methodology is that Nostraticists insist proto-languages be reconstructed as the basis for comparison, a procedure that Greenberg skips.

To English speakers, it may not be instantly obvious that their language has anything whatsoever in common with Finnish, Turkish, or Inuit, let alone Japanese, as the Eurasiatic hypothesis asserts. Given the speed of language change, and the 10,000 years or more that separate all these daughter tongues from the assumed proto-Eurasiatic, only a few echoes would be expected. As Greenberg's critics rightly point out, it is hard to be sure that the signal of these faint echoes rises above the noise of chance resemblance.

But consider the comparison of English with, say, Japanese. Given that *wakaru* means understand in Japanese, guess the meaning of *wakaranai*. Apart from the oddity of putting a negative at the end of the verb, it seems natural that *wakaranai* should mean don't understand, and so it does.

In many Indo-European languages, questions are expressed with words starting with "k" or "kw" sounds, though the "kw" has become a "w" in English. French has *quoi* (what?), Italian *come* (how?) and Latin *quando, quis*, and *quid pro quo*. So *wakaranaika*? Don't you understand?

It could be just by chance that the Indo-European and Japanese families use "k" sounds for question words. But an interrogative in *k* is found in every branch of Eurasiatic, Greenberg says.[296] In the Uralic family, Finnish has *ken*, meaning who? In Altaic, Turkic has *kim*, with the same meaning. In all dialects of Eskimo who is *kina*.

There are many interrogative words, so if one rummages around in all the languages of a proposed family, it's perhaps not so hard to find a few k-words. The same may be true of n-words for negatives. Greenberg's case for Eurasiatic rests not on any specific case but on the combination of a large number of such similarities that he has turned up. These include 72 types of grammatical similarity, though most are shared by only some of the eight postulated families of Eurasiatic. Nonetheless, "This grammatical evidence is quite sufficient in itself to establish the validity of the Eurasiatic family," Greenberg says.

Turning to words, as distinct from grammar, it's probably reasonable to assume that a given sound will ricochet around a related set of meanings over time. The assumption raises the chances of spotting a relationship between language families, but also of picking up accidental similarities. No single group of cognate words is conclusive, but large numbers can begin to make a case. Greenberg has found 437 groups of cognates for Eurasiatic, though very few have examples from every family.[297] One of the most interesting concerns a set of meanings based on the putative Eurasiatic word for finger, which Greenberg thinks was *tik*. Raise your first or index *tik* and you make a universally understood sign for the number one. Point it horizontally and you are drawing attention to something. On that basis, Greenberg cites the following echoes of this ancient word.

In the Indo-European family, linguists have reconstructed a proto-Indo-European root *deik*, meaning to show, from which comes the Latin word *digitus* for finger, and the English words digit and digital. In the Altaic family, the Turkish word for sole or only is *tek*. In the Korean-Japanese-Ainu group, there is Ainu's *tek* and Japanese's *te*, both meaning hand. As for Eskimo-Aleut, Greenlandic has *tikiq* for index finger, Sirenik and Central Alaskan Yupik have *tekeq*.

Greenberg put particular emphasis on another group of cognates, which he saw as providing a link between Eurasiatic and Amerind. It is a set of meanings centered on the word hand and including both give (to give is to hand something over) and measure (the width of the hand is often used as a measure, and in English is the name of the unit for measuring the shoulder height of horses and ponies). Many American Indian languages use a *ma* or *mi* sound as their word for hand (Algonquian *mi*, Uto-Aztecan *ma*, Tequistlatec *mane*, Guato *mara*). In the Eurasiatic family, Indo-European has a root *me-* meaning measure, whence metric, as well as Latin *manus*, hand; Gilyak has *man*, to measure by hand spans and *-ma*, a word added to numbers to indicate units of hand spans; Korean has *mān*, an amount or measure.

In Greenberg's view, Eurasiatic and Amerind were sister superfamilies, younger than the original languages of the Old World, of which the strange isolate languages like Basque and Burushaski (spoken in a small region of the northwestern Indian subcontinent) are relics. "The Eurasiatic-Amerind family represents a relatively recent expansion (circa 15,000 [years before the present]) into territory opened up by the melting of the Arctic ice cap. Eurasiatic-

Amerind stands apart from the other families of the Old World, among which the differences are much greater and represent deeper chronological groupings," he wrote in his last work.[298] It was, perhaps, a final gibe at his critics, who insisted that languages could be traced back no farther than 5,000 years or so; Greenberg was insisting he could see three times farther than they.*[299]

Echoes of the First Language

Nothing makes linguists heave wearier sighs than talk of the ancestral human language. The subject, in their general view, is not worth even talking about because, as every serious specialist knows, the roots of language cannot be traced back farther than 5,000 years, 10,000 at the very most. "Given present knowledge of language change and probability," writes Johanna Nichols, ". . . descent and reconstruction will never be traceable beyond approximately 10,000 years. Methods now being developed reach back much earlier but do not trace descent. Among other things, this means that linguistics will never be able to apply phylogenetic analysis to the question of when language arose and whether all the world's languages are descended from a single ancestor."[300]

Though Nichols's prediction may prove correct, biologists are not quite so pessimistic. With DNA, their phylogenetic trees reach back hundreds of millions of years, and 50,000 years ago is like yesterday. If Indo-European started to split up 8,700 years ago, as Gray's statistics say, languages may be reconstructible far further back in time than linguists have supposed.

---

* Greenberg died at the age of 85, the year after publishing the first volume of his work on Eurasiatic. Paul Newman, a linguist at Indiana University, recalls visiting him in his final illness. Greenberg mentioned his regret that he had not gotten around to classifying the languages of southeast Asia. "He looked at me," Newman said, "almost with tears in his eyes, and said that without classifying them, he hadn't finished his work with the world's languages."

Another friend, Harold Fleming of Boston University, a specialist in African languages, paid a farewell visit together with Greenberg's longtime colleague Merritt Ruhlen. The two were the last scholars to talk with him. The conversation turned to a recent work on Gilyak, an obscure language of northern Sakhalin. Fleming mentioned that there was a Gilyak word *irf* meaning fox or jackal. Greenberg was familiar with the word and noted that its "r" was trilled like a French "r."

"What a memory!" Fleming wrote in a tribute. "What a scholar! And what a shame that his vast and unique knowledge of human languages had to leave us, could not be electronically stored, and that such a great scientist had to die under a cloud of misguided criticism!" Source is note 299.

The very existence of Swadesh lists is proof that some words are retained longer than others. Might some be retained for long enough to reconstruct the tree of language 5 times farther back than Gray has done, close to the source of the ancestral tongue? Some words—*new, tongue, where, thou, one, what, name, how*—have half-lives greater than 13,000 years, and another seven words—*I, we, who, two, three, four, five*—are even more resistant to change, according to calculations by Mark Pagel. Such words, in his view, "can potentially resolve very old time depths," beyond the 5,000 to 10,000 years so often proposed for linguistic data.[301] The word for one, he notes, has a half-life of 21,000 years. This means it has a 22% chance of not changing in 50,000 years.

Could these long-lived Swadesh words support a genealogy that coalesced on a single proto-language? Greenberg played with the idea that he had found a word that might be a remnant of the mother tongue. It is the group of cognates, mentioned above, that are based on the set of ideas one/ finger/point and derived from the root *tik. Greenberg spotted what he assumed were cognates of this word in at least one member language of many of his language superfamilies. He mentioned the group in a lecture in 1977 but never published it, whether because of his own reservations or from fear of incurring more than the usual deluge of ridicule from his fellow linguists.

In the Eurasiatic family, as noted above, *tik words range from the English digital and Greek *daktulos* to Eskimo *tiqik* for "index finger." According to Ruhlen, Greenberg first noticed when defining the Nilo-Saharan family that several of its languages had words of the general form *t-k* for the word one.[302] The word for one in proto-Afroasiatic has been reconstructed *tak. In the Austroasiatic family, Cambodian or Khmer has *tai* as the word for hand, and Vietnamese has *tay*. In Amerind languages there are several *tik*-like words meaning finger or alone.

Even linguists who support Greenberg have little patience with the suggestion that the *tik word may be an echo of the mother tongue. Yet given Pagel's calculations, it is not impossible that some words still spoken today have very ancient pedigrees, and even that Greenberg's *tik is indeed a faint but indelible whisper from the distant days when the world was one.

11

HISTORY

The remarkable success of the English as colonists, compared to other European nations, has been ascribed to their "daring and persistent energy"; a result which is well illustrated by comparing the progress of the Canadians of English and French extraction; but who can say how the English gained their energy? There is apparently much truth in the belief that the wonderful progress of the United States, as well as the character of the people, are the results of natural selection; for the more energetic, restless, and courageous men from all parts of Europe have emigrated during the last ten or twelve generations to that great country, and have there succeeded best. . . . Obscure as is the problem of the advance of civilisation, we can at least see that a nation which produced during a lengthened period the greatest number of highly intellectual, energetic, brave, patriotic, and benevolent men, would generally prevail over less favoured nations. Natural selection follows from the struggle for existence; and this from a rapid rate of increase. It is impossible not to regret bitterly, but whether wisely is another question, the rate at which man tends to increase; for this leads in barbarous tribes to infanticide and many other evils, and in civilised nations to abject poverty, celibacy, and to the late marriages of the prudent. But as man suffers from the same physical evils as the lower animals, he has no right to expect an immunity from the evils consequent on the struggle for existence. Had he not been subjected during primeval times to natural selection, assuredly he would never have attained to his present rank.

CHARLES DARWIN, THE DESCENT OF MAN

WITH SETTLEMENT and the invention of agriculture, human societies embarked on a trajectory quite different from the foraging life that had hitherto been their only choice. The new behaviors that had now been developed allowed people to construct complex societies and urban civilizations.

They learned to treat strangers as kin, at least in the context of reciprocal exchanges and trade. They coordinated their activities through religious
rites. They defended their territory against neighboring tribes, or attacked
them when the moment seemed propitious. With settlement came specialization of roles, administrators to take control of surpluses, priests to organ-
ize religious ceremonies, headmen and kings to manage trade and defense.

The first cities started springing up in southern Mesopotamia some
6,000 years ago. Uruk, in what is now Iraq, sprawled over some 200 hectares
(500 acres) with large public buildings. The city required armies of laborers
and an administration to recruit and feed them. As societies became more
intricate, their operation demanded more sophisticated skills and perhaps
more specialized cognitive abilities, ones at least that no forager had had occasion to exercise. The invention of writing around 3400 BC opened the way
to the beginning of recorded history. The first great urban civilizations
emerged in Egypt, Mesopotamia, India and China. The next phase of the
human experiment had begun.

Genetics, which illumines many aspects of prehistory, yields even greater
returns when applied to the historical past because it can be related to
known people or events. DNA can be used to analyze populations, saying
who came from where, which helps understand mixtures of people like those
of the British Isles. DNA faithfully records who slept with whom throughout
the ages, a matter of historical interest in cases like the secret family of
Thomas Jefferson. And with populations that have married within themselves for centuries, like those of Jews, DNA can reach back to the time of
the patriarchs.

Geneticists may in future be able to trace back human lineages or pedigrees to all times and places, providing a genetic framework for exploring almost every historical period. Meanwhile a promising start has been made, as
is evident from the following cases.

The Secret Strategy of Genghis Khan

In the year 1227 the Mongol conqueror Genghis Khan died, perhaps in a fall
from his favorite horse. His empire stretched from the Caspian Sea to the
Pacific Ocean and included much of Russia, China, and Central Asia. His

followers brought his body home to a hill in northeast Mongolia. To keep his burial place secret, all those who interred him are said to have been killed, and their assassins were dispatched in turn.

Whether or not that story is true, Genghis's tomb remains secret and has defied two recent attempts, one by a Japanese expedition, one by Americans, to locate it. But while archaeologists were frustrated in their search for Genghis's hoard, geneticists engaged on a quite different task stumbled across a more vital part of Genghis Khan's legacy.

A team led by Chris Tyler-Smith of Oxford University had analyzed the Y chromosomes of some 2,000 men from populations across the Eurasian land mass. They noticed that many of the chromosomes fell into a single cluster. Some chromosomes in the cluster were identical at each of 15 sites tested and others were just one mutational step removed from this master sequence. The striking feature of the cluster was that the owners of its Y chromosomes did not all come from a single population, as would have been expected, but from regions all over Eurasia.

A clue to their origin was that the Y chromosome with the master sequence was particularly common in Inner Mongolia. A quarter of the men tested from this region carried the master sequence chromosome or its close derivatives. Another clue was that only 16 of the 50 or so Asian populations studied included men with the master sequence, yet all but one of these 16 live within what were the borders of the Mongol empire at the time of Genghis's death. The one exception was the Hazara of Afghanistan and Pakistan, who are thought to be descendants of Mongol soldiers sent to garrison the region.

Tyler-Smith and his colleagues believe the master sequence chromosome must be that of the Mongol royal house. It would have been carried by Genghis Khan and by the male relatives he sent to administer the regions of his far flung empire. Dating methods suggest the cluster started to form around 1,000 years ago, the time that Genghis's dynasty began its ascent to power.[303]

Mongol soldiers doubtless raped many women during their extraordinarily cruel and murderous campaigns. But there may be a more significant reason for the existence of so many men carrying the specific chromosome of the Mongol royal house: Genghis accumulated a large harem in which he seems to have labored with surprising industry. The fourteenth century Persian historian Rashid ad-Din, who served as chief minister of the Mongol government of Persia, wrote that Genghis Khan had nearly 500 wives and concubines, and that it was his practice to take women into his harem as booty whenever he conquered a new tribe.

Another Persian historian of the Mongol empire, 'Ata-Malik Juvaini, includes without further explanation the following observation in his *History of the World Conqueror*, completed in AD 1260: "Of the issue of the race and lineage of Chingiz-Khan there are now living in the comfort of wealth and affluence more than 20,000. More than this I will not say but shall rather avoid [the subject] lest the readers of this history should accuse the writer of these lines of exaggeration and hyperbole and ask how from the loins of one man there could spring in so short a time so great a progeny."[304]

Genghis's interest in procreation was shared by his sons, one of whom is credited with 40 sons. It seems to have been a deliberate policy of Genghis and his heirs to father as many children as possible. "It's pretty clear what they were doing when they were not fighting," comments a historian of the Mongol period, David Morgan of the University of Wisconsin.[305]

From the proportion of Mongol royal house Y chromosomes in their sample, Tyler-Smith and his colleagues have been able to calculate just how well Genghis succeeded in his procreative program. An astonishing 8% of males throughout the former lands of the Mongol empire carry the Y chromosome of Genghis Khan. This amounts to a total of 16 million men, or about 0.5% of the world's total.

The second most common Y chromosome in East Asia, after that of Genghis Khan, is one that probably belongs to Giocangga, the patriarch of the Manchu rulers who governed China as the Qing dynasty from 1644 to 1912. The Qing imperial nobility consisted of male descendants of Giocangga and his grandson Nurhaci, who founded the dynasty. The nobility was highly privileged and its members were able to keep many concubines. In addition the Qing nobility used marriages to cement political alliances with other peoples of northern China such as the Mongols.

Tyler-Smith has detected the Manchu chromosome in 7 northern populations though not in the Han, the major Chinese ethnic group. He believes the chromosome belongs to the Manchu royal house because of its frequency, its geographical distribution and the fact that its founder, according to genetic evidence from the chromosome itself, lived some 500 years ago—

Giocangga died in 1582. Tyler-Smith estimates that the Manchu Y chromosome is carried by 1.6 million men living today.[306]

A third patriarch, one with an estimated 2 to 3 million living descendants, has come to light through a study of Irish Y chromosomes. He may well be Niall of the Nine Hostages, an Irish high king of the fifth century AD whom some historians had regarded as a probably legendary figure.[307]

The genome offers a unique new window into history, one that is especially illuminating when DNA evidence can be combined with historical evidence. The cases of Genghis, Giocangga, and Niall of the Nine Hostages raise the question of whether large-scale procreation isn't just a perk of political power but may be a salient, even if unconscious, motivation for it.

A History of Britain, from the Genome's Viewpoint

The genome often holds surprising answers for historical questions that involve lineages. Consider the matter of English surnames. Commoners acquired surnames between AD 1250 and 1350, apparently for the convenience of feudal record keepers who needed to differentiate between tenant farmers with the same first names. The surnames were not highly original. They tended to be a person's profession (Smith, Butcher), or a patronymic (Johnson, Peterson), or derived from some landscape feature (Hill, Bush). Historians assumed that the same name had been invented many times over, so there would be no reason to assume that people with the same surname had a common ancestor in the thirteenth century. George Redmonds, however, a historian of British surnames and place names, came to feel that many English surnames had single progenitors. "But it was never possible to prove it genealogically because we don't have enough evidence," he says.[308]

That began to change when Redmonds's advice was sought by Bryan Sykes, a geneticist at Oxford University. Sykes had been invited to give a talk to scientists at Glaxo Wellcome, a large British drug company, which in the mid-1990's was beginning to take an interest in the human genome. The organizers of the conference at which Sykes was to speak asked him several times if he was related to the company's then chairman, Sir Richard Sykes.

He kept saying no, not that he knew of. Even the company chauffeur who arrived to drive him to the conference asked the same question.

Sykes was about to repeat his usual denial but suddenly a thought crossed his mind. "Maybe Sir Richard and I were related after all, *but without realizing it*," he writes. "And, more to the point, maybe I could prove it by a genetic test." Sykes asked the chauffeur to wait and rushed back to his lab for a genetic sampling kit (essentially a swab to brush cells off the inside of the cheek). At the conference, which his namesake was attending, he asked him for a sample.[309]

The two men had grown up in quite different parts of the country. Bryan Sykes's family lived in Hampshire, in southern England, Richard Sykes had grown up in Yorkshire, in the north. Apart from both having been trained as scientists, they seemingly had nothing else in common. But it turned out there was something else: they possessed the same Y chromosome.

Y chromosomes, of course, are bequeathed from father to son just as are surnames. After the test with his namesake, Bryan Sykes wondered if other Sykeses too might be related to one another. Research showed that there were many Sykeses living in Yorkshire and that the surname itself was derived from a Yorkshire word, *sike*, meaning a moorland stream. Sykes picked some 250 of his namesakes at random from the Yorkshire area and sent each a letter asking for a sample of his DNA. About a quarter obliged, and from analysis of their cells a distinct pattern emerged. About a half carried the identical Y chromosome, or one that was just a single mutational step away from it. The rest had a miscellany of unrelated Y chromosomes.

Several interesting conclusions followed. First, there was just one real Sykes Y chromosome. All the men who carried it were presumably descended from the first bearer of the surname. That meant the surname had been assigned only once or, if more than once, all other lines had ended without male heirs and no longer existed.

As for the 50% of Sykeses who did not carry the true Sykes Y chromosome, their cases must have been largely the result of what geneticists delicately refer to as a "nonpaternity event" at some point in their family tree, meaning the biological father was not the same as the father of record. Adoption is one possible explanation for nonpaternity, though it probably wouldn't account for many cases.

If half of Sykes men alive today have a nonpaternity event somewhere in their genealogy, doesn't that raise considerable doubt about the virtue of Sykes wives through the ages? Bryan Sykes argues this is not the case. Assuming there have been 23 generations of Sykes since the first Mr. Sykes in the thirteenth century, an infidelity rate of merely 1.3% per generation would account for the fact that only half of contemporary Sykes men carry the correct Y chromosome. This compares very favorably with the nonpaternity rates of contemporary populations, Sykes comments, which run from 1.4% to 30%, though most fall in the 2 to 5% range.[310]

"I'm proud to say I have the aboriginal chromosome," Sykes replied when asked whether he was a true Sykes or one of the out-of-wedlock kind. His early ancestors seem to have been a rough lot; they appear regularly in court records of the fourteenth century as having incurred fines for cutting down trees or stealing sheep. "Nonetheless, their wives were faithful through all this," Sykes says.[311]

Redmonds, the local historian, has traced the earliest Sykeses to the villages of Flockton, Slaithwaite and Saddleworth in West Yorkshire. The first mention of the name is a court record of AD 1286 referring to a Henri del Sike of Flockton. There are still Sykeses living in Flockton. Redmonds was able to locate the plot of land of which Henri del Sike was tenant, a farm that straddled a stream between two parishes. He took his geneticist friend to visit. "There was no sign of the farmhouse which my ancestor, the very first Sykes, had occupied, but even so, it felt quite extraordinary to be here," Sykes wrote. "Looking round at the old mill, the track and the stream, it seemed that nothing in the landscape had greatly changed. Nor had it. The field and croft boundaries were as they had been in the late thirteenth century when Henri del Sike was living here. As I stood, I could almost hear the voices of the children—my ancestors—laughing as they threw pebbles into the stream."[312]

Sykes has analyzed the Y chromosomes of three other English surnames and found that, as with his own, each can be traced to a single bearer. Because of his research it now seems that many English surnames once had a single bearer, and even the commonest ones like Clark and Smith may be descended from only a few originals.

Genetic analysis is at the least a new tool for historians and may one day support a new kind of history, possibly somewhat at variance with the conventional kind. English schoolchildren are taught that their history really begins with the Roman invasion of 55 BC and Caesar's defeat of the Celtic tribes who opposed him. The true bearers of the English heritage, the textbooks imply, are the Anglo-Saxons, later invaders whose Germanic language was the ancestor of English. The defeated Celtic inhabitants of Britain are assumed to have been pushed back into the hinterlands of Wales and Scotland and largely disappear from most history books.

But a survey of British Y chromosomes shows that the Y chromosomes characteristic of Celtic speakers, far from having disappeared, are carried by a large proportion of the male population of Britain. Nowhere does the indigenous population seem to have been wiped out, either by the Anglo-Saxons who invaded from Denmark and northern Germany in the sixth and seventh centuries AD, or by the Danish and Norwegian Vikings who arrived in the ninth and tenth centuries.[313] (Two other groups of invaders, the Romans and the Normans, probably arrived in numbers too small to have left a demographic mark.)

The Y chromosomes common among Celts have a particular set of DNA markers known to geneticists as the Atlantic modal haplotype, or AMH. AMH Y chromosomes are also found, it so happens, in the Basque region of Spain, whose inhabitants are thought to represent the original inhabitants of Europe. AMH–type Y chromosomes are particularly common in places like Castlerea in central Ireland, which no invaders ever reached. This suggests that the chromosomes are the signature of the first hunter-gatherers who arrived in Britain and Ireland toward the end of the Pleistocene ice age 10,000 years ago.

Given the similarity between Basque and Irish Y chromosomes, some geneticists suspect that people who had used Spain as a southern refuge during the Last Glacial Maximum started to move northward as the glaciers melted. Many may have traveled by boat up the west coast of Europe, entered the waterway between Ireland and England and settled on each side of it.[314]

The carriers of the AMH Y chromosomes presumably spoke a language like Basque or some other tongue belonging to the first Paleolithic inhabitants of Europe. So it is a puzzle that the chromosome is now associated with Celtic, an Indo-European language that spread to Britain only in the first century BC, along with ironworking technology and agriculture. The solution is presumably that the Celtic way of life became widespread in Britain mostly by cultural transmission, not by a large invasion of Celts. The cultural shift evidently included the adoption of Celtic language by the original inhabitants of the British Isles.

Another layer in this puzzle is that British mitochondrial DNA—the genetic element inherited solely through the female line—shows a different pattern from the Y chromosomes. The mitochondrial DNA generally resembles that of northern Europe. This suggests that the Celtic speakers in Britain obtained many of their wives from northern Europe, perhaps in exchange with European Celts, perhaps by pillage and rapine.[315]

The historian Norman Davies opens his recent history of the British Isles by noting that the mitochondrial DNA recovered from bones buried some 8,000 years ago in a cave in the Cheddar Gorge matched that of a local schoolmaster, proving the continuity of the human population of the region.[316] The genome is already being welcomed by historians as a rich new source of unexpected information.

The Origin of Icelanders

England was invaded by Vikings from both Denmark and Norway. The influence of the Danish Vikings can be seen most strongly in Y chromosomes from York and Norfolk in the eastern regions that bore the brunt of the Danish invasions. The Norwegian Vikings operated to the north of the Danes. In the ninth century AD they captured the Orkney Islands to the northeast of Scotland and made them a base of operations. Norn, a form of old Norse, was spoken on the islands until the eighteenth century. Norwegian Vikings have left a strong genetic signature among Orcadians, as Orkney Islanders are known, but their traces can also be seen farther afield, particularly in Iceland.

From their base in Orkney the Norwegian Vikings sailed around the northern coast of Scotland and down the waterway between Britain and Ireland, making settlements on both the British and Irish sides. In AD 870, the Vikings discovered Iceland, several days' sail to the northwest of Scotland. Apart from some Irish hermits, who quickly left, the island was uninhabited. News of this virgin territory, with no hostile natives, soon got around, and for 60 years there was a steady stream of settlers. Immigration ceased in AD 930, perhaps because many of the trees had been chopped down, prompting an ecological crisis, and there was no more unclaimed farmland left. The island was then essentially closed to new immigration until modern times.

Iceland's genetic history has received much attention, both for its intrinsic interest as an isolated human population and because its population has become a leading source for discovering the genetic roots of common diseases ranging from cancer and heart disease to asthma and schizophrenia. These diseases are thought to result from several errant genes acting in combination. The errant genes are very hard to detect because each makes only a small contribution to the overall disease. For various reasons, including an excellent system of medical records, Iceland offers many advantages in searching for such genes. In 1996 Kari Stefansson, an American-trained Icelandic neurologist, put together a high powered genetic analysis company, DeCode Genetics, which has enjoyed considerable success in identifying disease genes in Icelanders and other populations. The company and its large pharmaceutical partners hope to develop diagnostic tests and drugs on the basis of the Icelandic findings. It is therefore of considerable interest to know if Icelanders are genetically similar enough to other populations, particularly those of the United States and Europe, for discoveries about their patterns of genetic disease to be relevant elsewhere in the world.

Icelandic records from the twelfth and thirteenth centuries, notably documents known as the Book of Settlements and the Book of Icelanders, indicate that although Norse Vikings directed the immigration to Iceland, the inflow included people from the Norse settlements in the Orkneys and the coastal regions of Scotland, northern England and Ireland. Most of these Norse invaders, after the initial conquest, had intermarried with the local population. Assuming these Vikings brought their families, many if not most of the women in the founding Icelandic population would have been British or Irish, and in either case of Celtic origin. The Book of Settlements mentions only a small proportion of the founding settlers by name but of those whose ancestry is recorded, only 5% of the men came from the British Isles but 17% of the women. In addition, the Vikings captured slaves in raids in both countries, many of whom were probably women.

Icelandic historians have developed the case that their country was probably founded by men who were mostly Norse and women who were mostly from the British Isles, especially Ireland. This claim of descent from two important peoples, the Vikings and the Celts, helped to differentiate the Icelanders of the nineteenth and twentieth centuries from their much-resented rulers, the Danes. "The result of this conflation is the dominant modern concept of Icelandic origins, one that fuses the nobility and heroism of the Norse with the literary and other cultural traditions of the Irish and other peoples of the 'Celtic fringe,'" write a group of Icelandic and other experts.[317]

Given the power of modern genetics to deconstruct complex populations like that of England, it should be a simple problem to analyze the genetics of Iceland and check the validity of the historians' position. But it's not so easy. Comparison of Icelandic Y chromosomes with those in Scandinavia and the British Isles confirms that most of the male founders were indeed Norse, though not overwhelmingly so: some 20 to 25% of Iceland's founding fathers appear to have had Gaelic, meaning Celtic, ancestry, with the rest being of Norse origin.[318]

The founding mothers are much harder to trace. The patterns of mitochondrial DNA found among Icelanders today look generically European but without greatly resembling those of any particular country.[319] They look, well, Icelandic. The reason is probably genetic drift, the random gain or loss of genetic signatures between generations, accentuated by the violent fluctuations in Iceland's population since the settlement. The Black Death killed 45% of the population in 1402–1404. A smallpox epidemic in 1708 reduced the population by 35%. Famine, the aftermath of a volcanic eruption, caused a 20% decline in 1784–1785. After each of these population declines and expansions, the characteristic mix of mitochondrial DNA signatures would have changed, pushing the population farther down a separate path from that of its source populations.

With the genetic answers being so Delphic, a team of researchers has resorted to the old technique of craniometry. They measured Icelandic skulls from the settlement period and compared them with medieval skull collections from Ireland and Norway. Unfortunately the Icelandic skulls were not in good enough condition to tell their sex. Overall, they seemed very similar to the Norse skulls and less like those of Ireland. The researchers say that "although our results do not preclude a significant Irish or other contingent among the settlers of Iceland, we conclude that the founding population was of predominantly (60–90%) Norwegian origin."[320]

To help in its quest for disease genes, DeCode Genetics has assembled a genealogical database of the Icelandic population that extends back 1,100 years into the past. It is based on calfskin documents that hold the first 300 years of records, on church archives, and on the data from three complete censuses that were held starting in 1703. DeCode's genealogist, Thordur Kristjansson, reckons the database includes the names of about half the Icelanders who have ever lived, including 85% of those born in the nineteenth century and almost everyone who lived in the twentieth century.[321] The database has enabled DeCode researchers to explore the historical dynamics of a human population in fascinating detail.

One finding is that generation times are shorter in mother-to-daughter lines of descent than in father-to-son lineages. The average interval between generations was 29 years in female lineages going back to 1698, 32 years for male lineages. The difference presumably reflects the simple fact that women tend to be younger than their husbands.

A greater surprise is how many people in one generation leave few or no descendants in the next. DeCode has traced the ancestry of all 131,060 Icelanders born from 1972 to 2002 back to two cohorts of ancestors. Of all contemporary Icelandic women born since 1972, 92% are descended from only 22% of the women born in the 1848–1892 cohort, and 86% of contemporary men are the progeny of just 26% of this group.

The progeny pyramid narrowed even more steeply going back to an earlier generation of ancestors, those born between 1698 and 1742. Because of the incompleteness of the genealogy for earlier centuries, the pedigree of many contemporary Icelanders could not be traced that far back. Nevertheless, DeCode researchers found that just 7% of the women born in the early eighteenth century period are the ancestresses of 62% of contemporary women, while 10% of the men of this period fathered 71% of contemporary males.[322] Most people, in other words, have lines of descent that eventually go extinct, at least in a population the size of Iceland's, while just a few ancestors give rise to the majority of subsequent population cohorts.

This difference in reproductive success seems to be due largely to genetic drift, the force of which depends on the size of the population. Iceland's population fell to a post-settlement low of 33,000 after the 1708 smallpox epidemic but steadily increased from the beginning of the nineteenth century to its present level of 290,000. Even during this expansion, the influence of genetic drift was still at work. Before the end of the Pleistocene, there may have been many human populations no bigger than this, offering much grist for drift to work on.

Jewish Origins

The population history of Jews has been studied more than that of most other groups and has yielded one surprise after another. The population's first remarkable feature, from which all the others follow, is that Jews have to a significant extent married among themselves over the centuries. Jewish communities, in other words, have been largely endogamous, at least until recent times, which means the population's gene pool has had time to develop its own private history, and this genetic history has shed light on many historical events.

An important consequence of endogamy is that the gene pool is not diluted through intermarriage and so the selective pressures that may act on a population are able, over time, to favor specific genetic variations. A striking possibility, plausible though not yet confirmed, is that one particular Jewish community, the Ashkenazim of northern and central Europe, lived for a long time under a harsh selective pressure that raised certain variant genes to high frequency. These variant genes are well known to physicians because of their serious side effects—when inherited from both parents they cause a variety of serious diseases. But the variant genes can hardly have become so common through their role in promoting disease. They must confer some special benefit, and that, the hypothesis goes, is increased intelligence.

The selective pressure, according to this idea, was the restriction of Ashkenazim by their European host populations to a small number of occupations that happened to require higher than usual mental agility. The pressure lasted from about AD 800 to 1700. If true, the hypothesis, described further below, has several interesting implications, including that it would represent a very recent and dynamic example of human evolutionary change.

Judaism is a religion, open to others to convert to, and it has long seemed that religion and culture, not necessarily genetics, were the common elements of and between the world's various Jewish communities. But in

2000 a team of geneticists led by Michael Hammer of the University of Arizona reported that men from many far flung Jewish communities have the same set of variations on their Y chromosomes. The variations are not exclusive to Jews but are common throughout the Middle East.[323] The finding meant that the founding fathers of Jewish communities around the world were drawn from the same ancestral Middle Eastern population of 4,000 years ago from which other peoples, such as Arabs, Turks and Armenians, are also descended. These generic Middle Eastern Y chromosomes, part of the J branch of the worldwide Y chromosome family tree, are both a common link between men of different Jewish communities and proof that their communities must have remained genetically separate from their non-Middle Eastern host populations.

But genetics points to a very different story with Jewish women. A team under David Goldstein of University College, London, surveyed Jewish communities of Germany and eastern Europe, known as Ashkenazi Jews, as well as those of Morocco, Iraq, Iran, Georgia, Bukhara, Yemen, Ethiopia and India. Unlike the case with the Y chromosome, they found that each Jewish community has its own pattern of mitochondrial DNA variations, evidence that Jewish women, unlike Jewish men, do not all come from the same ancestral population.

Mostly, the mitochondrial DNA in each Jewish community doesn't closely resemble that of any other population, meaning that the geographic origin of the founding mothers of Jewish communities cannot be identified for certain. However, in several cases it looks as if it could come from the host community. For example, among the Bene Israel, the Jewish community of Bombay in India, the commonest pattern of mitochondrial DNA is just one mutation away from a pattern common among non-Jewish Indians.

The explanation proposed by Goldstein and his colleagues is that the founding fathers of Jewish communities came from the Middle East, the founding mothers from the host population in each country.[324] The Jewish men, arriving perhaps as traders and presumably unmarried, took wives from the local population in each country, and it seems then converted their wives to Judaism. Once the community was established and reached sufficient size, it became closed; no more wives were taken from the host population, and community members married among themselves. With no fresh infusions from the local population, the mitochondrial DNA in each Jewish community fell under the influence of genetic drift, making it look less and less like that of the local version from which it originated.

If this explanation is correct, the members of a Jewish community are generally a genetic admixture between Middle Easterners (the founding fathers) and the host population of each country (the founding mothers). This could explain why Jews often resemble the people of their host country, yet also in some respects resemble one another.

The genetic findings are generally compatible with Jewish historical accounts, though not in every detail. The ancestral Jewish population is ancient but came from a mix of Middle Eastern men, DNA analysis indicates, not a single patriarch. Many Jewish communities have accounts or traditions of how they were founded, often to escape persecution or at the invitation of a friendly potentate. The Iraqi Jewish community (whose members now live mostly in Israel) is said to have been founded after the destruction of the first temple in 586 BC. The Bene Israel of Bombay say their ancestors fled to India to escape the persecution of Antiochus Epiphanus, who ruled from 175 to 163 BC. The DNA analysis in general confirms that Jewish communities are ancient, though it cannot place an exact date on their founding. But the circumstance it suggests for their origin, that of single Jewish men taking local wives, indicates that at least some Jewish communities probably began as trading outposts, not by the mass emigration of families.

The modern Jewish population falls into three main groups, based on ancestral place of origin. Ashkenazi Jews lived mostly in Germany and eastern Europe and, from at least the sixth century AD, spoke a common language, Yiddish; Sephardic Jews are those expelled from Spain and Portugal in AD 1492 during the Spanish Inquisition; and Oriental Jews are those who have always lived in the Near East. Of the 5.7 million Jews living in the United States, some 90% are of Ashkenazic origin; of the 4.7 million Jews in Israel, 47% are Ashkenazic, 30% Sephardic and 23% Oriental.[325]

Jewish status, except for converts, is now defined by maternal descent. This practice, however, goes back only to Talmudic times, the period from around 200 BC to AD 500. In ancient Israel, tribal affiliation was determined by patrilineal descent, as were the two castes of hereditary priests, the cohens and the levites. After the destruction of the temple, the cohens were left with little to do and power passed into the hands of the rabbinate. The rabbis established matrilineal descent as the basis of Jewish identity. It is sometimes suggested they did so in wise appreciation of the fact that maternal descent is a fact and paternal descent only a probability; but a modern scholar, Shaye Cohen of Harvard University, believes rabbinic tradition and the influence of Roman law are likelier reasons.[326]

The patrilineal priestly tradition still exists, and has afforded geneticists another deep insight into Jewish history. Cohens and levites continue to carry out ceremonial roles in certain congregations. Cohens are called first to the reading of the Torah in synagogue, and are asked on special occasions to bless the congregation. (The cohen's blessing, signaled by holding up the hand with a split between the middle and the ring fingers, is familiar to many non-Jews; it was adapted by Leonard Nimoy, who remembered seeing it as a boy in synagogue, as the Vulcan greeting for his role as Spock in *Star Trek*.)[327]

Oral tradition holds that all cohens, or cohanim, are descended from Aaron, the brother of Moses and the first high priest. The Jewish priesthood is thought to have been established some 3,300 years ago and to have passed from father to son ever since. This fact was on the mind of Karl Skorecki, a medical researcher at the Technion-Israel Institute of Technology in Haifa, one morning when he was sitting in synagogue and the Torah was being read. The cohen doing the first reading was a Sephardic Jew. Skorecki, whose family is Ashkenazic, himself comes from a line of cohanim. The thought occurred to him that though he and the Sephardi differed strongly in physical appearance, they must both have inherited the same Y chromosome from Aaron, if oral tradition was correct.[328]

Skorecki called Michael Hammer, the University of Arizona geneticist, who agreed with his inference and set about analyzing the Y chromosomes of cohanim from both the Ashkenazic and Sephardic communities. Despite the millennium or so for which the two communities have been separate, and the geographical distance between them, Hammer and his colleagues found that the cohanim of both groups did indeed possess a distinctive genetic signature.

The signature is a set of DNA sequences at two specific sites on the Y chromosome. It is known as the cohen modal haplotype, a geneticist's phrase meaning the set of DNA variations typical of cohens. The Hammer team detected the cohen modal haplotype in 45% of Ashkenazic cohanim and in 70% of Sephardic cohanim.[329] The finding substantially confirmed the oral tradition that cohanim are descended from a single individual. This person was presumably a founding high priest and could perhaps have been Aaron himself if indeed there was an Aaron; some modern scholars believe the great patriarchs of Israel may have been more a part of legend than of history.[330]

To learn more about when the ancestor of all the cohanim might have lived, another team of geneticists including Skorecki and David Goldstein has looked at the variations that have developed on the cohen modal haplotype. The Goldstein team estimates that about 106 generations must have occurred to account for the observed amount of variation that has built up on the cohen modal haplotype. Assuming 30 years per generation, this means the ancestor of the cohanim lived some 3,180 years ago (or 2,650 years ago, if a generation time of 25 years is preferred).[331] A general date of about 3,000 years ago is of particular interest since it would place the first cohen at the beginning of First Temple Period of Jewish history.

The fact that only 50% or so of cohens, depending on the population, carry the cohen Y chromosome means that the rest must result from a discrepancy, at some point in their lineage, between the biological father and the father of record. Adoption cannot be invoked since the priesthood cannot be transferred to adopted sons, which leaves infidelity as the explanation. But as with the case of the English Sykeses, it takes only a small rate of nonpaternity in each generation to produce a large proportion of males with discrepant paternity many generations later.

Since the cohen lineage stretches back three times as far as that of the Sykeses, the fidelity of cohen wives must have been even higher. James Boster, an anthropologist at the University of Connecticut, calculates on the basis of the Skorecki team's figures that the rate of nonpaternity was 1.2% per generation among Ashkenazic cohanim and 0.4% among Sephardic cohanim. (This estimate would of course not pick up any cases where a cohen's wife had taken another cohen as her lover.)

Such infidelity rates are extremely low compared with the nonpaternity rates of 5% and more that are assumed typical of contemporary Western societies. Boster and his colleagues ask how cohanim through the ages secured such exemplary fidelity from their wives without resorting to the coercive measures used by men in other societies, such as purdah or chastity belts. They point to Jewish law and custom, under which intercourse is regarded as ritually impure from the beginning of a woman's menstruation until seven days after its end, whereupon it is the husband's duty to make love to her. Indeed he must do so immediately on her return from the ritual cleansing bath. This sage religious obligation has a strong consequence on the biological plane: it ensures that first intercourse, after several days abstinence, coincides with the three day period of peak fertility prior to ovulation. "This practice, coupled with extreme sanctions against adultery, . . . could account for these very high degrees of paternity certainty," the researchers observe.[332]

Levites, according to their genetics, have a more complicated story. Levites are a junior priesthood to the cohaniṁ, with fewer duties and obligations. By tradition, levites consist of all male descendants of Levi who are not also cohanim. The exclusion arises because Levi, the third son of the patriarch Jacob, was also an ancestor of Aaron. About 4% of Jewish men are levites, the same proportion as are cohanim.

The Y chromosomes of Ashkenazic and Sephardic levites show no particular similarity. So, unlike the case with the cohanim, there is no identifiable male levite lineage that precedes the Ashkenazi-Sephardi split. There is, however, a strong genetic signature common to 52% of Ashkenazic levites. It is a set of genetic variations belonging to a branch of the world Y chromosome tree known as R1a1. To judge by the amount of variation on these levite R1a1 chromosomes, the original ancestor seems to have entered the Jewish community about 1,000 years ago, roughly the time when Jewish settlement in northwest Europe began, in other words at the founding of the Ashkenazic community.[333]

The geneticists who discovered the R1a1 signature among the levites, a team that included Skorecki, Hammer and Goldstein, note that outside the Jewish community the R1a1 chromosome is relatively common in the region north of Georgia, in the Caucasus, that was once occupied by the Khazar kingdom. The Khazars were a Turkic tribe whose king converted to Judaism in the eighth or ninth century AD.

The geneticists propose that one or more of the Khazar converts may have become levites, accounting for the R1a1 signature among today's Ashkenazic levites. But Shaye Cohen, an expert on Jewish religious history, believes it unlikely that converts would become levites, let alone founding members of the levite community in Europe. The Khazar connection is "all hypothesis," in his view.

The genetic findings about cohen and levite ancestry are just genetics; they have no bearing on who is or is not considered to be a cohen or a levite. "Genetics is not a reality under rabbinic law," Cohen observes.[334]

Ethnic origins and hereditary priesthoods have opened two windows on Jewish history; a third has been created by the study of genetic diseases. Every population has its own particular set of genetic diseases, but those of Jewish communities in the United States and Israel have come under particular medical scrutiny, which is one reason why so many have been documented. The diseases are known as Mendelian, because they are caused by a single mutation and inherited in an obvious pattern; this stands in contrast with the so-called complex diseases, like cancer or diabetes, which can be caused by many contributing genes and are not inherited in any clear pedigree.

So far at least 40 different Mendelian diseases have been detected in Jewish populations.[335] Some of these diseases occur in non-Jewish populations as well, some are common to several Jewish communities, and some are restricted just to the Jews of a single community. The diseases are of course studied so as to help the patients but incidentally they yield many interesting clues to population history.

A disease called familial Mediterranean fever is caused by an errant gene that occurs among Ashkenazi, Iraqi and Moroccan Jews. It is also found in Armenians, the Muslim Druze sect and Turks. All present versions of the gene seem to be descended from a single ancestor who must have lived about 4,000 years ago in the ancient Middle Eastern population from which Jews and other ethnic groups are descended.

Later, the Jewish religion was founded and its adherents developed their own genetic history as they started to marry among themselves. The Jewish population may have grown to about a million people before suffering a terrible decline in AD 70, the year of the destruction of the temple in Jerusalem by a Roman army. That event began the diaspora, the dispersal of Jewish populations around the Mediterranean world. The largest Jewish community, the Ashkenazim of central and eastern Europe, may have reached 150,000 or so people by AD 1095, the year of the first crusade and the beginning of the persecution of Jews by Christians.

The Ashkenazi Jewish population is of particular interest because it has produced many individuals of high intellectual achievement, both in Europe and among the Ashkenazim who fled to the United States and elsewhere in the wake of Nazi persecution. Another attribute is a distinctive set of Mendelian diseases. The mutations that cause these diseases can hit at random anywhere in the genome, so would not be expected to favor any particular category of gene. But no fewer than four of the Ashkenazic Mendelian diseases affect the cell's management of chemicals known as sphingolipids, so called because their discoverer could not resolve the sphinxlike riddle of what they did. The four sphingolipid diseases are Tay-Sachs, Gaucher, Niemann-Pick and mucolipidosis type IV. Another cluster of four diseases affects the cell's system for repairing DNA. These are the BRCA1 and BRCA2 mutations which can cause breast and ovarian cancer, Fanconi's anemia Type C and Bloom syndrome.

The sphingolipid diseases in particular are reminiscent of the group of mutations that cause blood disorders like sickle cell anemia, and which are now recognized as defenses against malaria. When malaria suddenly became a threat some 5,000 years ago, natural selection favored any mutation that offered protection, even if it carried serious disadvantages. Diseases like sickle cell anemia are the result of that quick fix. The sickle cell mutation, though devastating for individuals unlucky enough to inherit a copy from each parent, offers substantial protection against malaria for the much larger number in the population who inherit just a single copy.

Evolution has probably engineered many quick fixes like this in the human genome. Later, as the generations pass and better mutations turn up, evolution is generally able to improve on the quick fix or favor variant genes that diminish the side effects of the first mutation. This is why a batch of harmful mutations affecting a common pathway is the fingerprint of a recent evolutionary response to some sudden selective pressure.

Turning back to the four sphingolipid diseases, they look awfully like an evolutionary quick fix, a set of mutations selected because of some advantage gained by disrupting sphingolipid metabolism. So if that advantage was protection against disease, what disease could it have been? The puzzle is that carriers of the sphingolipid mutations don't seem to enjoy unusual immunity to any specific disease.

"A second hypothesis," writes Jared Diamond, after discussing the idea that the variant genes conferred greater resistance to tuberculosis, "is selection in Jews for the intelligence putatively required to survive recurrent persecution, and also to make a living by commerce, because Jews were barred from the agricultural jobs available to the non-Jewish population."[336]

The suggestion that one group of people may be genetically more intelligent than another is a sensitive subject, not least because it opens the door to the argument that if some groups are smarter, others may be less so. The idea Diamond floated was not followed up, and indeed the geneticists who next looked at the sphingolipid diseases suggested they had grown common not through natural selection but because of a quite different mechanism known as a founder effect.

If a population gets squeezed down to small numbers by some calamity, and then expands, its gene pool will be an amplified version of that of the few individuals who survived the disaster. If one of the survivors carried a generally rare mutation, the mutation will be much commoner in the new expanded population than it is in the general human population. The relatively high incidence of the usually rare mutation in the expanded population is called a founder effect, after the founder who carried the mutation.

Recently Neil Risch, now of the University of California, San Francisco, concluded that the four sphingolipid diseases must have become common among the Ashkenazi Jewish population because of founder effects. He noted that the four diseases had similar properties to the other Ashkenazic Mendelian diseases, such as having arisen very recently, in the last 1,100 years. Because all the Mendelian diseases seemed therefore to have arisen through the same cause, he argued, that cause must be founder effects, since natural selection wouldn't favor such a miscellany of different mutations.[337]

A similar conclusion was reached for different reasons by Montgomery Slatkin of the University of California, Berkeley.[338] He calculated that if there had been two bottlenecks in the Jewish population, at AD 70 following the destruction of the temple, and at sometime after AD 1100, the founder effects caused by these two population reductions could explain how the Ashkenazic disease genes had gotten to be so common. Slatkin's calculation did not rule out natural selection, but since a founder effect was possible, that provided the most economical explanation, in his view.

But a new and substantially buttressed case for natural selection, with the need for extra intelligence posited as the driving force, has now been advanced by Gregory Cochran, a physicist turned population geneticist, and Henry Harpending, an anthropologist at the University of Utah.[339] They agree with Risch that all the diseases arose from the same cause and at about the same time. But the cause must have been natural selection, not founder effects, because in testing other, non-disease causing Ashkenazic genes, Cochran and Harpending could see no evidence for any of the reductions in population size required to cause Risch's founder effects. Nor is there any clear historical evidence, they say, that the Ashkenazi Jewish population ever dwindled to the low numbers needed to generate a founder effect.

Having argued that natural selection must therefore have been the reason that the Ashkenazic mutations became so common, Cochran and Harpending next ruled out disease as the agent of selection. The Ashkenazic population, they note, lived in the same cities as their European hosts and suffered from the same diseases, yet Europeans show no similar pattern of mutations.

But there was a significant difference between Ashkenazim and Europeans, Cochran and Harpending argue, and it lay in the special range of occupations to which Ashkenazi Jews were restricted by their Christian hosts.

The origin of the Ashkenazi Jews is obscure but they were established in northern France by shortly after AD 900. Most had become moneylenders by AD 1100 because Christians forbade usury, and this continued for several centuries. Moneylending was an intellectually demanding profession, not least because the Indian numerals in use today, and specifically the concept of zero, did not become widespread in Europe until around 1500. Figuring out xvii percent of cccl, without the use of zero, is not a straightforward computation.

Jewish communities became subject to particular persecution after the First Crusade, launched in AD 1095. They were expelled from England in 1290, from France in 1394, and from various regions of Germany in the fifteenth century. Many migrated to Poland, where they lived first as moneylenders and then served as the managerial class for the Polish authorities, particularly in such roles as tax farming. (The tax farmer would pay a nobleman the tax due, then try to recoup the sum, with profit, from the peasantry.) Being frequently uprooted and forced to start over again, there was continual pressure on families to survive and find ways of being useful to their unpredictable hosts. "From roughly 800 AD to 1650 or 1700 AD, the great majority of the Ashkenazi Jews had managerial and financial jobs, jobs of high complexity, and were neither farmers nor craftsmen. In this they differed from all other settled peoples of which we have knowledge," Cochran and colleagues write.

Restrictions on Ashkenazi employment were lifted around 1700, bringing to an end a period of some 900 years during which most of the population would have had to earn a living in occupations requiring more mental ability than most. Given what is known about the heritability of intelligence, the Cochran team calculates that even in as little as 500 years there would have been time for the intelligence of the Ashkenazi population to have been raised appreciably.

The authors cite evidence suggesting that sphingolipid mutations serve to foster the growth and interconnectedness of neurons, sometimes by lifting natural restraints. They believe that all the Ashkenazic disease mutations, in ways that remain to be discovered, serve to promote the extra cognitive skills that the Ashkenazic population needed in order to survive.

The outcome, they say, is that Ashkenazim have an average IQ of 115, one standard deviation above that of northern Europeans, although some measurements put it at only half a standard deviation higher. This is the highest average IQ of any ethnic group for which reliable data exist. Such an advantage may not make much difference at the average, where most people are situated, but it translates into a significant difference at the extremes. The proportion of northern Europeans with IQs greater than 140 is 4 per thousand but the figure for Ashkenazim is 23 per thousand, a sixfold difference.

This may have something to do with the fact that Ashkenazim make up only 3% of the U.S. population but have won 27% of U.S. Nobel prizes. Ashkenazim account for more than half of world chess champions. "Jews and half Jews, who make up about 0.2 percent of the world's population, have won a total of 155 Nobel prizes in all fields, 117 in physics, chemistry and medicine," writes the anthropologist Melvin Konner.[340]

Jewish folklore holds that intelligence was fostered not by occupation but by channeling the cleverest children to become rabbis. The rabbis were able to have more children, the folklore explanation holds, because they were sought as husbands for the daughters of wealthy families. "Talmudic academies served as systems of selection," writes Konner. "Whatever we think of what was studied, the process culled the best minds in every generation of Jews for more than a thousand years. Rising stars among these bright young men would board with successful merchants, and matches would be made between them and the merchants' daughters."

But the Cochran team gives short shrift to this explanation, saying there were not enough rabbis—only 1% of the population—to make a genetically significant difference. As further proof of their thesis, they cite the fact that the two other main branches of the Jewish community, Oriental Jews and Sephardim, lived mostly under Muslim rulers who often forced them into menial jobs, not the intellect-demanding ones imposed on Ashkenazim. Oriental Jews and Sephardim score similarly to northern Europeans with no elevation in IQ, as would be predicted under the Cochran team's thesis.

Among Ashkenazim, some 15% carry one of the sphingolipid or DNA repair mutations, and up to 60% carry one or other of all the disease mutations. (Most of these diseases are only harmful if a mutated gene is inherited from both parents, and some others are not fully "penetrant," a geneticist's term meaning a person can carry the mutation but doesn't necessarily have the disease.)

In summary, the Cochran group has taken two well accepted phenomena—the odd pattern of Ashkenazic Mendelian diseases and the notable intellectual achievement of Ashkenazim—and has attempted to establish a link between them. The argument is necessarily extended, but is carefully developed at each stage. "It's certainly a thorough and well argued paper, not one that can easily be dismissed outright," said Steven Pinker, a cognitive scientist at Harvard. Though several aspects of the argument cross disputed academic territory—it assumes that intelligence is heritable and that IQ scores are a reliable measure of it—it has the virtue of making a clear and testable prediction: that people carrying one of the Ashkenazic mutations should do better than average on IQ tests. As of this writing, the test has not been conducted.

Despite the existence of genetic diseases that can be called Jewish, in the broader context Jews are doubtless highly similar to other populations in the west Eurasian or Caucasian branch of the human family. Their genesis as a distinctive group resembles that of Icelanders. Just as Jews appear to be a mixture of Middle Easterners with various European or other Middle Eastern populations, Icelanders also are probably a mix of two Caucasian populations, Norwegians and Celts. Both Jews and Icelanders have practiced endogamy, the necessary step for keeping one's gene pool to oneself, Jews for religious reasons, Icelanders for geographic ones. Icelanders have been genetically separate for 1,000 years, but are still so similar to other Europeans that they can serve as a test bed for discovering European disease genes; Jews have been separate for just 2,000 years longer.

DNA and the Secret Family of Thomas Jefferson

Leading American historians for years denied a startling circumstance that was clearly attested to in the historical record: Thomas Jefferson, the third president of the United States, fathered an unacknowledged family with his slave mistress Sally Hemings.

Here is some of the evidence that historians of Jefferson found reason to disbelieve:

"It is well known," the journalist James T. Callender wrote in the Richmond *Recorder* on September 1, 1802, the second year of Jefferson's first presidency, "that the man, *whom it delighteth the people to honor*, keeps, and for many years past has kept, as his concubine, one of his own slaves. Her name is SALLY. The name of her eldest son is TOM. His features are said to bear a striking although sable resemblance to those of the president himself."

In 1873 a son of Sally Hemings, Madison Hemings, gave a long biographical statement to an Ohio newspaper, the Pike County *Republican*. He told how his mother, then aged around 13, had been sent to Paris, where Jefferson, then a widower, was American ambassador. Sally's role was to be a servant to Maria, one of Jefferson's two daughters.

"Their stay (my mother and Maria's) was about eighteen months," Madison Hemings related. "But during that time my mother became Mr. Jefferson's concubine, and when he was called back home she was *enceinte* by him. He desired to bring my mother back to Virginia with him but she demurred. She was just beginning to understand the French language well, and in France she was free, while if she returned to Virginia she would be reenslaved. So she refused to return with him. To induce her to do so he promised her extraordinary privileges, and made a solemn pledge that her children should be freed at the age of twenty-one years. In consequence of this promise, on which she implicitly relied, she returned with him to Virginia.

"Soon after their arrival, she gave birth to a child, of whom Thomas Jefferson was the father. It lived but a short time. She gave birth to four others, and Jefferson was the father of them all. Their names were Beverly, Harriet, Madison (myself) and Eston—three sons and one daughter. We all became free, agreeably to the treaty entered into by our parents before we were born."

It is difficult to believe that a 68-year-old Ohio carpenter, as Madison Hemings then was, would be moved on chance encounter with a journalist to invent an account of such specificity and poignancy. It contained many details that could be independently checked. Jefferson freed very few slaves, but he let all of Sally's children go free. Winthrop Jordan, a historian at the University of Mississippi, documented in 1968 that Jefferson, despite his many absences from Monticello, was present at the time of conception of all Hemings's known children.

But apart from Jordan, who stated that a liaison between Jefferson and Hemings was a possibility, a long line of Jefferson historians dismissed Madison Hemings's account. Merrill Peterson, the first historian to give it scholarly study, conceded that Madison's recollection "checks remarkably well with the data accumulated by scholars on Jefferson's domestic life and the Monticello slaves." But he chose to reject its central claim, that Madison was Jefferson's son, with the defective argument that since Jefferson's enemies wanted the story to be true, it must be false. The Jefferson-Hemings liaison was a legend, he wrote, sustained by the hatred of the Federalists, the propaganda of the British, "the Negroes' pathetic wish for a little pride," and the cunning of slave auctioneers thinking they would "get a better price for a Jefferson than for a Jones." The "overwhelming evidence of Jefferson's domestic life refuted the legend," Peterson assured his readers in 1960.[341]

A third of a century later, the historian Joseph J. Ellis took the same position. He derided the story as a "piece of scandalous gossip" that had affixed itself to Jefferson's reputation "like a tin can that then rattled through the pages of history." Jefferson historians had no desire to know what might be in the tin can; they just wanted to boot it as far away as possible. "Within the community of Jefferson specialists, there seems to be a clear consensus that the story is almost certainly not true," Ellis wrote in 1996. "After five years mulling over the huge cache of evidence that does exist on the thought and character of the historical Jefferson, I have concluded that the likelihood of a liaison with Sally Hemings is remote."[342]

The community of Jefferson specialists found much more to their taste a self-serving story concocted by the Jefferson family to protect his reputation. Two of Jefferson's grandchildren put it about that Jefferson's nephews Peter and Samuel Carr were the fathers of the light-skinned slaves at Monticello. The Carrs were the sons of Jefferson's sister, which could explain why the young slaves so resembled Jefferson.

There the matter rested, so far as scholars were concerned, and might well have solidified into accepted fact, but for the trespass of two outsiders onto the historians' carefully groomed turf. An African American lawyer, Annette Gordon-Reed, weighed the same evidence available to the historians but came to the opposite conclusion. A Jefferson-Hemings liaison was very likely, she argued, though it could not be proved. That finding led her to a harsher, but not so unreasonable, judgment that "those who are considered Jefferson scholars have never made a serious and objective attempt to get the truth of this matter."[343] Gordon-Reed had no genetic evidence available to her; she simply interpreted the available historical evidence more skillfully than a generation of professional historians had done.

The second outsider to the issue was Eugene Foster, a pathologist who had recently retired from Tufts University to Charlottesville, Virginia. Foster had no particular interest in Jefferson, but Charlottesville is Jefferson country, and a friend asked him one day if DNA fingerprinting might shed any light on the Hemings issue. Foster decided it wouldn't—forensic DNA analysis can identify individuals and resolve paternity but can't reach back up a genealogical tree because of the shuffling of DNA between generations. Then he learned of work on the Y chromosome, which is passed unchanged from father to son except at its very tips, and realized it could hold the answer.

First Foster needed a sample of Jefferson's Y chromosome. Unfortunately Jefferson had no male descendants, but his paternal uncle Field Jefferson would have carried the same Y chromosome, assuming no illegitimacy in the Jefferson male line. With the help of Herbert Barger, a Jefferson family historian, Foster located 5 male descendants of Field Jefferson and wrote explaining his project and asking for a sample of their blood.

Next, he needed a Y chromosome of the Carr brothers, the leading suspects in the view of historians and the Jefferson family members. He obtained blood from three male-line descendants of the three sons of John Carr, the grandfather of Peter and Samuel.

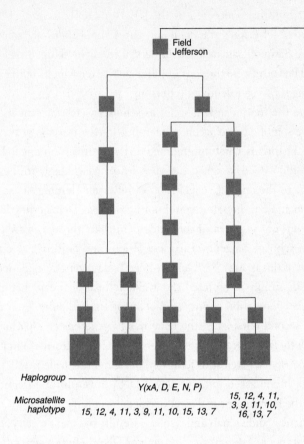

Haplogroup ——————————————————————————

Y(xA, D, E, N, P)

Microsatellite ——————————————————— 15, 12, 4, 11,
haplotype   3, 9, 11, 10,
15, 12, 4, 11, 3, 9, 11, 10, 15, 13, 7   16, 13, 7

FIGURE 11.1. THOMAS JEFFERSON'S FAMILY WITH SALLY HEMINGS.

A Y chromosome analysis of Thomas Jefferson's family performed by Eugene Foster and Chris Tyler-Smith showed that Eston Hemings, a son of the slave Sally Hemings, carried the same Y chromosome as that of Thomas Jefferson's male relatives and was therefore highly likely to have been Jefferson's son.

At specific sites on the chromosomes, short DNA sequences are repeated a number of times, as in ATATAT. The number of repeats changes quite often between generations, so can be used to identify different lineages. In this case the repeats at 11 sites have been used to fin-

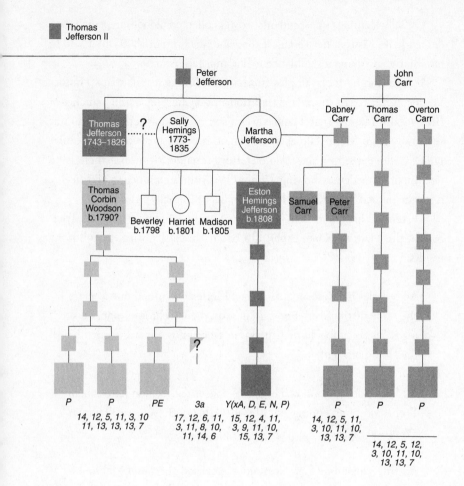

Thomas
Jefferson II

Peter
Jefferson

John
Carr

Dabney
Carr
Thomas
Carr
Overton
Carr

Thomas
Jefferson
1743–1826
?
Sally
Hemings
1773–
1835
Martha
Jefferson

Thomas
Corbin
Woodson
b.1790?
Beverley
b.1798
Harriet
b.1801
Madison
b.1805
Eston
Hemings
Jefferson
b.1808
Samuel
Carr
Peter
Carr

?

P
P
PE
3a
Y(xA, D, E, N, P)
P
P
P

14, 12, 5, 11, 3, 10
11, 13, 13, 13, 7

17, 12, 6, 11,
3, 11, 8, 10,
11, 14, 6

15, 12, 4, 11,
3, 9, 11, 10,
15, 13, 7

14, 12, 5, 11,
3, 10, 11, 10,
13, 13, 7

14, 12, 5, 12,
3, 10, 11, 10,
13, 13, 7

gerprint the different Y's. A natural shift in repeat numbers at the 9th site has occurred in the rightmost of Field Jefferson's descendants.

Eston Hemings had the same 11 repeats as the Field Jefferson descendants, so would have acquired his Y chromosome from a Jefferson family member, who from the historical evidence was almost certainly Thomas Jefferson, the third president of the United States. The finding strongly supports contemporary rumors that Jefferson had fathered a secret family with Sally Hemings, his slave mistress.

Of Sally Hemings's descendants, Foster collected blood from a male-line descendant of Eston, her youngest son, who was born in 1808 and is said to have borne a striking resemblance to Thomas Jefferson.

In addition he took samples from 5 male line descendants of Thomas Woodson. There is a strong oral tradition among long-separated branches of the Woodson family that Thomas was a son of Thomas Jefferson who was sent away from Monticello as a boy. There is one other reference, besides James Callender's, to a slave son of Jefferson named Tom. But there is no documentary evidence showing Thomas Woodson's presence at Monticello, nor is he named by Madison Hemings in his list of Sally's children.

Foster had his blood samples analyzed in the laboratory of Chris Tyler-Smith, the Y chromosome expert at Oxford University, with the following results:

- All 5 male-line descendants of Field Jefferson turned out to carry the same distinctive set of markings on their Y chromosome, making it highly probable that the same Y chromosome was carried by the third president.

- All 5 male descendants of Thomas Woodson carried non-Jeffersonian Y chromosomes, ruling out the idea that Jefferson was Thomas Woodson's father.

- All three male-line Carr descendants carried the same Y chromosome, proving this was the true Carr family Y chromosome.

- The Y chromosome of Eston Hemings's male-line descendant was a perfect match to the Jefferson family Y chromosome, and differed from that of the Carrs.

The "simplest and most probable explanations for our molecular findings," Foster and his colleagues wrote, "are that Thomas Jefferson, rather than one of the Carr brothers, was the father of Eston Hemings Jefferson, and that Thomas Woodson was not Thomas Jefferson's son." They could not rule out the possibility that some male Jefferson other than Thomas had fathered Eston, they wrote. "But in the absence of historical evidence to support such possibilities, we consider them to be unlikely."[344]

The DNA evidence by itself does not prove conclusively that Thomas Jefferson had an unacknowledged family with Sally Hemings. Nor does the historical evidence by itself. But for two entirely independent kinds of evidence to point so strongly to the same conclusion makes a robust case. The historian Joseph Ellis certainly felt so, and to his credit admitted error. "The new evidence persuaded me that I had been wrong, and I felt a kind of moral and professional obligation to say that," he said.[345]

Barger, the historian who helped Foster, was not pleased by this outcome. He has assailed Foster's findings and proposed other male Jeffersons as the father of Eston, just as the Jefferson grandchildren did on an earlier occasion. But none of these ad hoc candidates can be shown to have been present at Monticello at all of Sally Hemings's conceptions as Thomas Jefferson was.

Jefferson had a strange and special tie with Sally Hemings, one only possible in the divided world of slave and free. No portraits of Sally survive, but she may well have reminded Jefferson of his beloved wife Martha, being as she was Martha's half sister. Both were daughters of John Wayles, Martha by his wife Martha Eppes, Sally by the slave Elizabeth Hemings, who became Wayles's mistress after his wife's death.

Jefferson's feelings for Sally, a subject of much speculation, are simply unknown. In all his correspondence he mentions her just once. At Monticello his white family and his unacknowledged black family lived side by side, but even in private he seems to have paid no special attention to Sally's children. Madison learned to read, he says, "by inducing the white children to teach me the letters and something more." As for Jefferson, "He was not in the habit of showing partiality or fatherly affection to us children. We were the only children of his by a slave woman."

Some mysteries lie beyond the power even of DNA to resolve.

12

EVOLUTION

The main conclusion arrived at in this work, namely, that man is descended from some lowly organised form, will, I regret to think, be highly distasteful to many. But there can hardly be a doubt that we are descended from barbarians. The astonishment which I felt on first seeing a party of Fuegians on a wild and broken shore will never be forgotten by me, for the reflection at once rushed into my mind—such were our ancestors. These men were absolutely naked and bedaubed with paint, their long hair was tangled, their mouths frothed with excitement, and their expression was wild, startled, and distrustful. They possessed hardly any arts, and like wild animals lived on what they could catch; they had no government, and were merciless to every one not of their own small tribe. . . . Man may be excused for feeling some pride at having risen, though not through his own exertions, to the very summit of the organic scale; and the fact of his having thus risen, instead of having been aboriginally placed there, may give him hope for a still higher destiny in the distant future. But we are not here concerned with hopes or fears, only with the truth as far as our reason permits us to discover it; and I have given the evidence to the best of my ability. We must, however, acknowledge, as it seems to me, that man with all his noble qualities, with sympathy which feels for the most debased, with benevolence which extends not only to other men but to the humblest living creature, with his god-like intellect which has penetrated into the movements and constitution of the solar system—with all these exalted powers—Man still bears in his bodily frame the indelible stamp of his lowly origin.

CHARLES DARWIN, THE DESCENT OF MAN

LOOK BACK at the 5 million years since the human line split from that of apes. With the power of genetics, that story can now be told at a far deeper level of detail, abundantly confirming the extraordinary insight that Darwin first hinted at in his *On the Origin of Species* of 1859 and made more explicit in *The Descent of Man* in 1871. Humans are just one of the myriad branches of the tree of life, sharing the same fundamental genetic mechanisms as all other living species, and shaped by the same evolutionary forces. This is the truth, as far as our reason permits us to discover it. All differing accounts of human origin, though a matter of religious dogma for most of recorded history and widely believed to the present day, are myth.

Darwin's insight was the more remarkable because he had no concept of genes, let alone of DNA, the chemical script in which the genetic instructions are inscribed. Not until 1953 was DNA recognized to be the hereditary material and only since 2003 has the fully decoded script of the human genome been available for interpretation.

With this script in hand, we can begin to trace the finest workings of the grand process that Darwin could see only in outline. The picture is still far from complete. But as the previous chapters have recorded, a wealth of information has already been retrieved from the darkness. We can see how the human form was shaped, step by step, from the anatomy of an apelike forebear, losing its body hair and developing darker skin as recorded in the gene for skin color. Human behavior, whether in the search for reproductive advantage or the defense of territory, shows clear continuity with that of apes. But it also developed its own characteristic pattern with two pivotal steps: the emergence of long lasting bonds between men and women some 1.7 million years ago, and at 50,000 years ago the evolution of language. Language, a novel evolutionary faculty enabling individuals to share a sequence of precise thoughts symbolically, opened the door to a new level of social interaction. Early human groups developed the institutions that shape even the largest and most sophisticated of today's urban societies. These included organized warfare; reciprocity and altruism; exchange and trade; and religion. All were present in embryo in the hunter-gatherer societies of the Upper Paleolithic. But it required another development, a diminution of human aggression and probably the evolution of new cognitive faculties, for the first settlements to emerge, beginning 15,000 years ago, and it was in the context of settled societies that warfare, trade and religion attained new degrees of complexity and refinement.

Human nature is the set of adaptive behaviors that have evolved in the human genome for living in today's societies. We have developed, and can execute instinctively, the behaviors necessary for warfare, for trade and exchange, for helping others as if they were kin, for detecting outsiders and cheaters, and for immersing our independence in the religion of our community.

The narrative of the human genome explains our origins, our history, and our nature, but many of its implications are far from welcome to one group or another. "The human mind evolved to believe in the gods. It did not evolve to believe in biology," writes Edward O. Wilson.[346] Religion is not the only subject for which evolution provides a discordant view of the world. Geneticists are likely to provide ever greater detail about how individuals vary, how men and women have different interests and abilities, and how races differ. Scientists studying the genome may in time establish that many human motives, from mating behavior to traits of personality, are shaped by genetically based neural circuits, thus casting some doubt on the autonomy of human actions. But however discomforting such findings may be, to falter in scientific inquiry would be a retreat into darkness.

One of the most perplexing implications of Darwin's theory is that humans are the unplanned product of a blind and random process. Looking at our cousins, the chimpanzees, we seem so much more advanced than they, as if shaped for a higher purpose. This is in part an illusion our ancestors helped to create by eliminating all competing human species. The more deeply we understand chimpanzees, the more evident their similarities to people become. They are shaped from the identical clay, the gene pool of our common ancestor. Some 99% of their DNA sequence corresponds almost exactly to our own.[347] They are highly intelligent, feel empathy for others, fabricate a variety of tools, and lead a complex social life. But by chance and circumstance, chimpanzees took one path through evolutionary space, the human lineage took another. Perhaps the chimp path required rather little change, whereas the human lineage, seeking a way of life beyond the trees, became so different because it was constantly forced to innovate.

The relentless search for new solutions produced not one but a whole clutch of hominid species. At least three—the Neanderthals, *Homo erectus*, and *Homo floresiensis*—survived until modern humans made their exit from Africa. Had these archaic peoples endured till the present day, our own species would surely seem less special, being evidently just one of many ways in which evolution could spin variations out of the basic ape lineage.

But if evolution generates new species by mechanisms that are in part random, should human existence be ascribed just to a long sequence of chance events? In the passage quoted above, that concludes his *Descent of Man*, Darwin gives a typically careful answer, a yes with a reservation. Man has risen to the summit of the organic scale, he says, "though not through his own exertions." Yet "some pride" in the result would be excusable. Why so, if no human exertion was involved? The reference a few lines later to man's powers of sympathy, benevolence and intellect is presumably Darwin's answer, and we can perhaps begin at last to see what he meant.

Though evolution through natural selection depends on random processes, it is shaped by the environment in which each species struggles to survive. And for social species the most important feature of the environment is their own society. So to the extent that people have shaped their own society, they have determined the conditions of their own evolution.

The nature of this interaction between culture and evolution is not yet clear, because it has only just come to light. It has long been assumed by historians, archaeologists and social scientists that human evolution was completed in the distant past, probably before any kind of culture had begun, and that there has been no evolutionary change, or only a negligible amount, within the last 50,000 years or so. Even evolutionary psychologists, who are committed to explaining the mind in terms of what evolution shaped it to do, assume that evolution's work was completed in a preagricultural past more than 10,000 years ago.[348]

But the evidence now accumulating from the genome establishes that human evolution has continued throughout the last 50,000 years. The recent past, especially since the first settlements 15,000 years ago, is a time when human society has undergone extraordinary developments in complexity, creating many new environments and evolutionary pressures. Hitherto it has been assumed the human genome was fixed and could not respond to those pressures. It now appears the opposite is the case. The human genome has been in full flux all the time. Therefore it could and doubtless did adapt to changes in human society. And this may mean that people have adapted in various ways, both good and bad, to the kinds of society they lived in.

Following is a review of the evidence that evolution is an active and vigorous force in the human population, a brief look at some of the implications, and a discussion of where human evolution might be headed in the future.

Evolution in the Recent Human Past

At its most basic, the process of evolution is simply a change in gene frequencies between generations; one version of a gene, in other words, becomes more common in a population and other versions less common.[349] It may take many generations, however, for the shift in frequencies to become significant or for one version of the gene to supplant the alternative versions altogether. The belief that human evolution is essentially complete rests on the conjecture that it moves too slowly for significant changes to have occurred recently, for instance within the last 50,000 years.

But the human genome, just like those of all other species, can adapt quite quickly to changes in the environment. Without this capacity, humans would long ago have lapsed into extinction. At least four types of recent evolutionary change have already become evident and many more will doubtless come to light. The recent genetic changes already discovered include defenses against disease; increases in fertility; responses to cultural changes that affect the human environment; and changes in cognitive behavior.

Disease and other parasites are among the most serious threats to the welfare of large animals, and few diseases have presented a greater threat to human existence than malaria. Though the malaria parasite is very ancient, malaria is thought to have become a common disease among people only within the last 10,000 years, and perhaps within the last 5,000 years or so when slash-and-burn agriculture was introduced into West Africa. The sunlit pools in the clearings would have provided an ideal breeding place for the mosquitoes that carry the parasite.

Confronted with a severe and sudden threat like the *Plasmodium falciparum* form of malaria, natural selection will favor any helpful mutation that crops up. Several blood diseases, such as sickle cell anemia, have arisen because of changes in hemoglobin that protect against malaria. Another natural defense against the parasite is the impairment of an enzyme known as G6PD (for glucose-6-phosphate dehydrogenase) which kicks off the train of reactions leading to the metabolism of glucose. Genetic variations that sharply reduce the efficiency of the enzyme work wonders against the malaria parasite, although they also cause a serious blood disorder.

There are two principal variants of the G6PD gene, which seem to have arisen independently of each other. One is found in African populations, the other in peoples of the Mediterranean. Sarah Tishkoff of the University of Maryland has dated the age of the two G6PD variants. The statistical method she used gives a broad range of possible ages, but all are consistent with the idea that the human genome has evolved its resistance to malaria only recently. The variation of the G6PD gene that is common in Africa arose sometime between 4,000 and 12,000 years ago, according to Tishkoff's calculations. The Mediterranean variant started to spread between 2,000 and 7,000 years ago.[350]

An even more recent protection against disease evolved some 1,300 years ago among people of northern Europe. The protection consists of a modification to a protein, known as the CCR5 receptor, that is embedded in the surface of white blood cells. The variant gene, known as CCR5-delta-32 because it has lost 32 units from its DNA, occurs in some 14% of Swedes, diminishes to 5% in Mediterranean populations and is rare to nonexistent in non-European populations.[351] It presumably achieved its sudden prominence through the natural selection caused by some serious epidemic that raged through Europe at that time.

The variant gene was discovered because it is also protective against AIDS. But AIDS is far too recent a disease to have driven the frequency of the CCR5-delta-32 variant to such a high level in the European population. That driving force was at first thought to have been the Black Death, which killed some 25 to 40% of Europeans in the years 1346–1352, and a further 15 to 20% in 1665–1666. But it now seems that smallpox, which in the long term killed a larger number of people though in less dramatic fashion, was the probable agent of selection.[352] Presumably the loss of 32 units from the gene alters the structure of the surface protein used by both the AIDS virus and the smallpox virus to gain access to a cell.

Diseases, especially those that kill people before the age of reproduction, are potent selective forces. So too are genes that affect fertility. A genetic change that favors fertility has been observed at high frequency in European populations. It occurs rarely in Africans but bears the marks of being under strong selection in Europe, as if promoted by something in the European environment. The change consists of a large segment of chromosome 17, some 900,000 DNA units in length, which has become flipped, or inverted. The inversion carries several genes but it is not clear which of them is responsible for conferring greater fertility.[353] The inversion evidently rose to high frequency among Europeans after the exodus from Africa and perhaps in the last 10,000 years.

From a historical point of view, the most interesting class of evolutionary changes are those that have occurred in response to human culture. When people first started to abandon their way of life as hunters and gatherers some 15,000 years ago, they had much less need for two kinds of gene, the olfactory genes that mediate the sense of smell, and the genes that are used by the liver to detoxify the natural poisons with which wild plants defend themselves. When a gene is vital for an organism's survival, any mutation in the gene will be lethal and the mutated version will be lost from the population. But when mutations crop up in genes that don't matter anymore, the gene may survive, even though it has lost its function.

This has been the fate of many human olfactory genes. Mammals possess a standard suite of about 1,000 of these genes. The proteins that each makes are embedded in the surface of the cells that line the nose and serve to detect specific odors. Once people settled down and grew their food, they no longer depended on their noses to detect which fruits were ripe or which wild plants were relatively safe to eat. On evolution's use-it-or-lose-it principle, more than 60% of olfactory genes in people are now inactive. Yoav Gilad and colleagues at the Max Planck Institute for Evolutionary Anthropology in Leipzig find that humans are losing their olfactory receptor genes four times faster than other higher primates. "This process is probably still ongoing in humans," they conclude.[354] Distressing news as this may be to gourmets and oenophiles, the price of civilization is that the faculty of smell is inexorably being degraded.

In parallel with the loss of olfactory genes, people are also losing genes that detoxify natural plant poisons. The enzymes made by these genes are no longer needed for their original purpose but have assumed an unexpected role in modern societies—that of metabolizing medicinal drugs. This unnatural stimulus does not occur often enough, however, and many of the genes are being lost through disuse. (This process explains much of the variability in the response to drugs, including why some people have severe side effects or require different doses. People who have lost the gene that breaks down a certain drug will maintain a high dose of it in their bloodstream, whereas those who still retain the gene will clear the drug rapidly.)

After settlement and agriculture came the rearing of livestock. Lactose tolerance, as discussed earlier, is the genetic response to the availability of animal milk. The genetic change evolved some 6,000 years ago among cattle herders of northern Europe and later among peoples of Africa and the Near East who took up pastoralism.

Of two recent evolutionary cognitive changes, which are also responses to culture, one is the postulated development of genes for enhanced intelligence among the Ashkenazi Jews of medieval Europe. As noted in the previous chapter, the hypothesis holds that because of the intellectually demanding occupations to which Jews were confined for some 900 years, any mutation that released constraints on the growth of the brain was favored, and that these mutations are the ones familiar as causing a variety of genetic diseases among people of Ashkenazic descent.

The other cognitive change is one that can be inferred from the striking recent rise to prominence of versions of two brain genes. As described in chapter 5, the genes first came to light because mutated versions cause microcephaly, a condition in which people are born with an unusually small head and brain. The new version of the microcephalin gene appeared around 37,000 years ago, rapidly became more common under intense selective pressure, and is now carried by most people in Europe and East Asia.[355] The other gene, a new version of ASPM, emerged 6,000 years ago and is now carried by 44% of Caucasians.[356] Both genes are thought to be involved in determining the number of neurons formed in the cerebral cortex in the early embryo. The rapid spread of these two alleles indicates that the human brain has been subject to intense evolutionary pressures in the recent past and may still be evolving.

These instances of recent evolutionary change are probably just the first of many that remain to be discovered. Since all occurred after the dispersal from Africa, the alleles that cause them are present to a different extent in different populations. Human diversity therefore cannot be a purely cultural phenomenon, as many social scientists sometimes seem to believe. It has a genetic component too. The component remains to be defined and quantified, but it could prove to be substantial.

Evolution in History

Given these new examples of evolutionary change, it seems clear that human evolution has continued at the very least until the recent past, nor is there any reason to think it will ever cease. An obvious but far-reaching conclusion follows. Evolution and history are not two distinct processes, with one following another like the change between royal dynasties. Rather, evolution and history overlap, with the historical period being overlaid on a still continuing process of evolutionary change.

The implications are clearly of possible interest to historians and social scientists. Historians are concerned with motivation, but seldom consider sexual selection as a driver of national politics. Students of the Mongol empire have proposed many sophisticated reasons for the expansion of the Mongol empire, such as Genghis's supposed desire to prevent any future group of steppe dwellers rising to power in the way the Mongols had done. The discovery that 8% of Asian men in the lands ruled by Genghis Khan carry the Y chromosome of the Mongol royal house offers a quite different motive, but one of unusual specificity.

Sexual selection, in this case the effort by one male to propagate his genes at the expense of others, has been a powerful force throughout primate history, from chimpanzee societies to those of the Yanomamo. It has operated with little change in more complex societies, especially during times when access to women was one of the accepted rewards of power. Even in many contemporary societies, where at least a pretense of monogamy is expected of rulers, the old instincts have not disappeared. True, procreation played no evident role in the drive to power of dictators like Hitler or Stalin. But Mao Tse-tung, as revealed in the memoir by his personal physician Li Zhisui, lived like an emperor, with villas and swimming pools and a stream of girls procured by the Cultural Work Troupe of the Central Garrison Corps. "He was happiest and most satisfied when he had several young women simultaneously sharing his bed," writes his unadmiring Boswell.[357]

Presidents John F. Kennedy and Bill Clinton conducted affairs while serving in the White House. "I don't know a single head of state who hasn't yielded to some kind of carnal temptation, small or large. That in itself is reason to govern," said François Mitterrand, the French president whose funeral in 1996 was attended by both his wife and his mistress.[358] Yet the drive for reproductive success is not a motive cited in many histories. Perhaps the possibility that a brute desire to procreate might drive the affairs of state is a concept that historians find too gross to contemplate.

Given that physical characteristics, such as the ability to digest lactose, have evolved in recent history, so too may have many other traits, including changes in social behavior. At least two conditions are necessary for the human genome to be significantly modified: there must be a selective pressure applied steadily for several generations, and those who adapt to the pressure must have more descendants than others. Such conditions may have occurred quite often in the human past, although it is hard at present to identify them.

Even evolutionary changes need not be permanent. The aggressiveness of the Yanomamo could have a lot to do with the marginal nature of the environment in which some of them live. Under conditions in which aggressive men have more children, genes that favor aggression would become more common. If the Yanomamo should suddenly become peaceful traders for many generations, then a new set of genes might be favored. The fierce Vikings of the tenth century became the peaceful Scandinavians of today. A cultural explanation is usually taken for granted, on the assumption that genes cannot change so quickly. But maybe the speed with which natural selection can act in human populations has been underestimated. Biologists are only just beginning to understand the genes that affect social behavior, some 30 of which have so far been detected, mostly in various species of laboratory animal. One of the most interesting findings is of a genetic mechanism for bringing about quick evolutionary change in a gene for behavior.*[359]

A possible subject of future inquiry is whether longstanding traits of certain societies may have an evolutionary basis, perhaps because over many generations they allowed people with a certain kind of personality to enjoy

---

*The gene in question promotes good parental behavior in male prairie voles. Stuck in front of the gene is a section of DNA that changes length quite readily between generations. In a vole population the section exists in a spectrum of lengths. Males with the longer section look after their pups with devotion, males with the shorter sections are less attentive. In environments where good parenting pays off, males with the longer section will be more successful. But in environments where it does not, males with the shorter gene may predominate. The gene in question is called the vasopressin receptor gene. Humans possess the same gene and the variable section, but its effects in people remains to be understood. Source is note 359.

greater reproductive success than others. Some scholars have remarked on long term cultural differences between societies of East and West. Richard E. Nisbett, a social psychologist at the University of Michigan, believes there are "dramatic differences in the nature of Asian and European thought processes," principally that Westerners view the behavior of physical objects and organisms as being governed by precise rules, whereas East Asians seek to understand events in terms of the complex web of interrelationships in which they are embedded. The social structures of Europe and China are built to match, in Nisbett's view, with Asian societies being interdependent and Western societies individualistic.[360]

Another scholar, the military historian Victor Davis Hanson, attributes the continuous prowess of Western militaries since the era of classical Greece some 2,500 years ago to democratic institutions and willingness of the free yeomanry to accept effective military discipline while retaining their independence and initiative. "Western ideas of freedom, originating from the early Hellenic concept of politics as consensual government and from an open economy . . . were to play a role at nearly every engagement in which Western soldiers fought," Hanson writes.[361]

To the extent that such long term cultural traits indeed exist, what might be their origin? Nisbett cites the fact that Chinese civilization was founded on rice farming, which required irrigation and central control; hence ordinary Chinese found themselves living in a world of complex social constraints, whereas the ecology of ancient Greece favored activities like hunting, herding, fishing and trade, which could be pursued without an elaborate social organization.

Did rice farming encourage the conformity for which eastern societies are known and small-scale farming the rugged individualism of the west? Given the propensity of the human genome to adjust to its environment, including the social environment, it is not impossible that many societies have left their imprint in the genetics of their members, and that the character of different societies reflects the personality traits of those who were the most reproductively successful in them. This is perhaps what Darwin had in mind in allowing that people might take some credit for their evolutionary progress.

The extent to which such a process may have happened in history cannot yet be determined. The novel issue of recent human evolutionary change is of particular interest, however, because it bears on the question of which future directions human evolution is likely to take. As Darwin noted, the fact that man has evolved to his present state "may give him hope for a still higher destiny in the distant future."

Future Directions of Human Evolution

The most improbable feature of science fiction movies is not the faster-than-light travel or the transporter beams but a feature that audiences accept without a second thought: the people. The inhabitants of the far future are always portrayed as looking and behaving exactly like people today.

All that is certain about future evolution is that people will not remain the same as they are today. They will differ because the environments of the future will be different, and people will either have adapted to them or perished.

Many futures are possible. Perhaps the gracilization of the human form will continue as societies favoring trade and cooperation extend their advantage over those more inclined to aggression. Perhaps our distant descendants will be far more intelligent, having evolved in response to the ever increasing intellectual demands of a more complex society. Perhaps they will be stockier, with shorter arms and legs, having followed the standard biological rules for adapting to cold climates as the Earth plunges back again into the inevitable next ice age. Perhaps the human lineage will resume its speciation, dividing into two or more castes inhabiting different social niches. And if self-sustaining populations are ever established on Mars or Europa, they will certainly follow independent evolutionary paths, adapting in form and social behavior to the ecological rules of their new planet.

Theorists have not yet reached agreement on the future evolution of the human species. Two of the founders of population genetics, Ronald Fisher and Sewall Wright, disagreed strongly on the conditions that favor evolutionary change. Wright believed that evolution worked best when a population was divided into small independent groups with limited gene flow between them; a genetic innovation that emerged in one group could then be allowed to spread to the others. Fisher, on the other hand, thought beneficial innovations were more likely to arise in large populations with a high degree of mixing. "Which of them is right? No one really knows," says Alan Rogers, a population geneticist at the University of Utah.

As it happens, the conditions favored by Wright apply well to most of recent human history before 10,000 years ago, when the population was divided into small, distant groups spread across the globe. But the world at present, with increasing travel and migrations, seems much closer to Fisher's ideal conditions for evolutionary change. "You used to marry a lass from your local village, now it's anyone you can track down on the internet," says Mark Pagel, an evolutionary biologist at the University of Reading in England.[362]

In Fisher's major work, *The Genetical Theory of Natural Selection*, published in 1929, he developed the argument that genes for mental ability are more frequent among the wealthy, who have fewer children, whereas the poor, who tend to be less intelligent, have more children; therefore natural selection acts against genes that promote intelligence. This aspect of Fisher's work is not much discussed, because it was used to support the disastrous eugenics policies of the early twentieth century. But in the view of some population geneticists, its theoretical argument has not been refuted: at least in developed countries, people of higher intelligence tend to have fewer children, so it would seem that their genes cannot become more common in the next generation. Others argue that the poor tend to have more children from lack of education, not any lack of intelligence. "Fisher's empirical observation is correct, that the lower orders have more babies, but that doesn't mean their genotypes are inferior," says Pagel.

Human brain size and intelligence have clearly expanded throughout most of evolution, and it would be strange if this trend should suddenly grind to a halt just as societies, and the skills needed to flourish in them, have become more complex than ever. It would be stranger still if humans, selected throughout evolution on the basis of maximum fitness, the propensity to leave as many descendants as possible, should suddenly abandon this deeply ingrained behavior. Nor is there any evidence from IQ tests to suppose that human cognitive ability is falling, as Fisher predicted. Therefore, despite the apparent correctness of Fisher's premise, that in modern societies the rich and more intelligent tend to have fewer children, his conclusion of inexorable intellectual decline seems somehow to be false.

The reason, evolutionary psychologists suggest, is that the rich are able to invest more in their children—a college education makes a big difference to future success—and thus they may leave more descendants in the long run, even if they have fewer children. The argument assumes that children who are well educated and well endowed will have children of similar quality, generation after generation, whereas at a certain level of destitution fertility will be reduced. So at some level of wealth, the better way for parents to maximize their Darwinian fitness will be to have fewer children in the expectation of leaving many more great-grandchildren.

Whether this is the case in practice is unclear. Teasing out the relationship between wealth and fertility is no easy matter and the demographic data needed to resolve the issue seem to be lacking. "This is a tricky and subtle business," writes the evolutionary psychologist Bobbi Low, "and most currently available data, gathered to answer other questions, are inadequate."[363]

One way in which future human evolution will differ from that of the past is that in larger populations the effect of genetic drift is much diminished. The larger the population, the longer it takes for one version of a gene to supplant all the alternative versions. Since drift is a principal mechanism for reducing the diversity that is constantly introduced by mutation, it follows that human genomes will become more diverse as neutral mutations accumulate. Too much diversity, according to theoretical calculations, could eventually make people infertile unless they mated only with people whose genomes were similar to their own.[364] This would make it impossible for all humans to interbreed, as is the case at present, and confine people to seeking partners within genetically similar groups. Such an outcome would be another step in fragmenting the human population into different species.

The weakening of drift and its mutation-reducing effect might be offset, to some extent, by human intervention, in the form of genetic engineering. Biologists may soon learn how to modify eggs, sperm or the early embryo so as to insert corrective genes that remedy future health defects. New genes inserted into the human genome on a widescale basis to replace existing genes might have the same mutation-shedding effect as genetic drift.

Suppose the genetic modification eventually takes the form of adding many new genes, packaged in the form of an extra chromosome that could be introduced into a couple's eggs and sperm prior to an in vitro fertilization procedure, which a few decades hence has supplanted the quaint and hazardous method of conceiving at random.

This extra chromosome would include a suite of genes for correcting all genetic diseases diagnosed in the prospective parents. It would carry genes to fortify the immune system, to fend off cancer and to combat the cruel degenerative diseases of age. The in vitro fertilization procedure and the individually tailored genetic engineering would be expensive, but critics who claim that only the rich will benefit might be confounded should governments find the procedure to be so cheap, compared with the lifetime of health care costs it averts, that they offer it free to all citizens.

The early versions of the extra chromosome, to continue the scenario a little further, are allowed only to carry genes that correct threats to health. But when the first generation of humans to carry twenty-four pairs of chromosomes turns out to be entirely normal and robustly healthy, various enhancements of desirable qualities are allowed. The extra chromosomes carry genes that promote longevity, improve the symmetry and beauty of the body, and enhance intelligence, though all within carefully prescribed limits. After various adjustments, the technology is brought to a high level of perfection. The only downside is that the people with twenty-four chromosome pairs cannot interbreed with those carrying the old fashioned number unless the latter agree to genetic modification, which many resist. Once again speciation, the division of the human population into two or more species, is the unintended outcome.

Two choices lie ahead. One is between directed human evolution and the natural kind, the other is whether to allow or promote speciation. The idea of directing human evolution, by modifying the germline, may seem adventurous, but evolution's method rests on the outcome of two chance-driven processes, mutation and drift. It could be contended that despite the madness of its method, evolution has not done too badly so far. But evolution works with glacial speed. With germline modification, on the other hand, just as in the breeding of domesticated animals, human intervention can reach a desired outcome much more quickly.

The most serious disadvantage of actively managing the human germline probably lies in the risks incurred by unintentionally suppressing evolution's vast capacity for novelty. By creating mutations at random, and testing each out to see if it works, evolution comes up with innovations that no one would think of. Those in charge of modifying the human germline, on the other hand, doubtless constrained by medical ethics to avoid all risk, would inevitably freight the genome with their conservative preferences.

Speciation, the other major issue in the human evolutionary future, is another powerful way of generating novelty and hence of improving the essentially unfavorable odds that the human species will last a long time. Our previous reaction to kindred species was to exterminate them, but we have mellowed a lot in the last 50,000 years. A bifurcation into land people and sea people—mammals have returned to the sea several times already—might not necessarily lead to conflict, nor would that of separately evolving populations on Mars and Earth. More problematic would be different human species occupying the same environment, especially if one were somehow deemed inferior or bound in helotry to the other.

There is no one human evolutionary future but many possible paths, some to be shaped by chance, some by choice. We have come so far. There is so much farther to go.

Notes

1. Genesis 3: 7, 21.
2. Ralf Kittler, Manfred Kayser, and Mark Stoneking, "Molecular Evolution of *Pediculus Humanus* and the Origin of Clothing," *Current Biology* 13:1414–1417 (2003). Other researchers have challenged a technical aspect of the paper. The challenge, if sustained, would suggest a considerably older date, perhaps up until 500,000 years ago, for the evolution of the body louse and the invention of clothing. David L. Reed et al., "Genetic Analysis of Lice Supports Direct Contact between Modern and Archaic Humans," *Public Library of Science Biology* 2:1972–1983 (2004); Nicholas Wade, "What a Story Lice Can Tell," *New York Times*, October 5, 2004, p. F1.
3. Feng-Chi Chen and Wen-Hsiung Li, "Genomic Divergences between Humans and Other Hominoids and the Effective Population Size of the Common Ancestor of Humans and Chimpanzees," *American Journal of Human Genetics* 68:444–456 (2001).
4. Pascal Gagneux et al., "Mitochondrial Sequences Show Diverse Evolutionary Histories of African Hominoids," *Proceedings of the National Academy of Sciences* 96:5077–5082 (1999).
5. Richard G. Klein, *The Human Career*, 2nd edition, University of Chicago Press, 1999, p. 251. Unless otherwise specified, the paleoanthropological and archaeological facts in this chapter are mostly drawn either from this broad and lucid textbook, or from a more popular book that is based on it, *The Dawn of Human Culture* by Richard G. Klein and Blake Edgar, John Wiley & Sons, 2002.
6. Chen and Li, "Genomic Divergences." The number of DNA differences between two species depends on the size of the parent population and the number of generations for which the two species have been separate. If the generation time and the number of years since the split are known, geneticists can estimate what they call the "effective" population size. This is a theoretical population, which must be multiplied by a factor of two to five to get the census-size population.
7. P. S. Rodman, in *Adaptations for Foraging in Non-human Primates*, Columbia University Press 1984, pp. 134–160, cited in Robert Foley, *Humans before Humanity*, Blackwell, 1995, p. 140.
8. Richard G. Klein, *The Human Career*, 2nd edition, University of Chicago Press, 1999, figure 8.3, p. 580.

9. Roger Lewin and Robert A. Foley, *Principles of Human Evolution*, 2nd ed., Blackwell, 2004, p. 450.

10. Robert Foley, *Humans before Humanity*, Blackwell, 1995, p. 170.

11. Richard G. Klein, *The Human Career*, p. 292.

12. Richard Wrangham, "Out of the Pan, Into the Fire," in Frans B. M. DeWaal, ed., *Tree of Origin*, Harvard University Press, 2001, p. 137.

13. Richard G. Klein and Blake Edgar, *The Dawn of Human Culture*, Wiley, 2002, p. 100; Robert A. Foley, "Evolutionary Perspectives," in W. G. Runciman, ed., *The Origin of Human Social Institutions*, Oxford University Press, 2001, pp. 171–196.

14. Richard G. Klein, *The Human Career*, p. 292.

15. Charles Darwin, *The Descent of Man and Selection in Relation to Sex*, 2nd edition, 1874, p. 58.

16. Mark Pagel and Walter Bodmer, "A Naked Ape Would Have Fewer Parasites," *Proceedings of the Royal Society* B (Suppl.) 270:S117–S119 (2003).

17. Rosalind M. Harding et al., "Evidence for Variable Selective Pressures at MC1R," *American Journal of Human Genetics* 66:1351–1361 (2000).

18. Nina G. Jablonski and George Chaplin, "The Evolution of Human Skin Coloration," *Journal of Human Evolution* 39:57–106 (2000).

19. Alan R. Rogers, David Iltis, and Stephen Wooding, "Genetic Variation at the MC1R Locus and the Time Since Loss of Human Body Hair," *Current Anthropology* 45:105–108 (2004).

20. Nina G. Jablonski and George Chaplin, "Skin," *Scientific American* 74:72–79 (2002).

21. Arthur H. Neufeld and Glenn C. Conroy, "Human Head Hair Is Not Fur," *Evolutionary Anthropology* 13:89 (2004); B. Thierry, "Hair Grows to Be Cut," *Evolutionary Anthropology* 14:5 (2005); Alison Jolly, "Hair Signals," *Evolutionary Anthropology* 14:5 (2005).

22. Hermelita Winter et al., "Human Type I Hair Keratin Pseudogene phihHaA Has Functional Orthologs in the Chimpanzee and Gorilla: Evidence for Recent Inactivation of the Human Gene After the Pan-Homo Divergence." *Human Genetics* 108:37–42 (2001).

23. R. X. Zhu et al., "New Evidence on the Earliest Human Presence at High Northern Latitudes in Northeast Asia," *Nature* 431:559–562 (2004).

24. Robert Foley, *Humans before Humanity*, Blackwell, 1995, p. 75.

25. Richard G. Klein, "Archeology and the Evolution of Human Behavior," *Evolutionary Anthropology* 9(1):17–36 (2000).

26. Richard Klein and Blake Edgar, *The Dawn of Human Culture*, p. 192.

27. Richard Klein, *The Human Career*, p. 512.

28. Ian McDougall et al., "Stratigraphic Placement and Age of Modern Humans from Kibish, Ethiopia," *Nature* 433:733–736 (2005).

29. From a list of fifteen modern behaviors described by Paul Mellars in "The Impossible Coincidence—A Single-Species Model for the Origins of Modern Human Behavior in Europe," *Evolutionary Anthropology* 14:12–27 (2005).

30. Richard Klein, *The Human Career*, p. 492.

31. Sally McBrearty and Alison S. Brooks, "The Revolution That Wasn't: A New Interpretation of the Origin of Modern Human Behavior," *Journal of Human Evolution* 39:453–563 (2000).

32. Christopher Henshilwood et al., "Middle Stone Age Shell Beads from South Africa," *Science* 304:404 (2004).

33. James M. Bowler et al., "New Ages for Human Occupation and Climatic Change at Lake Mungo, Australia," *Nature* 421:837–840 (2003).

34. Michael C. Corballis, "From Hand to Mouth: The Gestural Origins of Language," in Morten H. Christiansen and Simon Kirby, *Language Evolution*, Oxford University Press, 2003, pp. 201–219.

35. Marc D. Hauser, *The Evolution of Communication*, MIT Press, 1996, p. 309.
36. Ibid., p. 38.
37. Steven Pinker, *The Language Instinct*, William Morrow, 1994, p. 339.
38. Marc D. Hauser, Noam Chomsky, and W. Tecumseh Fitch, "The Faculty of Language: What Is It, Who Has It, and How Did It Evolve?" *Science* 298:1569–1579 (2002).
39. Derek Bickerton, *Language and Species*, University of Chicago Press, 1990, p. 105.
40. Ray Jackendoff, *Foundations of Language*, Oxford University Press, 2002, p. 233.
41. Frederick J. Newmeyer, "What Can the Field of Linguistics Tell Us about the Origins of Language?" in *Language Evolution*, by Morten H. Christiansen and Simon Kirby, Oxford University Press, 2003, p. 60.
42. Nicholas Wade, "Early Voices: The Leap of Language," *New York Times*, July 15, 2003, p. F1.
43. Steven Pinker, e-mail, June 16, 2003, quoted in part in Wade, "Early Voices."
44. Steven Pinker and Paul Bloom, "Natural Language and Natural Selection," *Behavioral and Brain Sciences* 13(4):707–784 (1990).
45. Derek Bickerton, *Language and Species*, p. 116.
46. Ray Jackendoff, *Foundations of Language*, p. 240.
47. Ann Senghas, Sogaro Kita, and Asli Özyürek, "Children Creating Core Properties of Language: Evidence from an Emerging Sign Language in Nicaragua," *Science* 305:1779–1782 (2004).
48. Wendy Sandler et al., "The Emergence of Grammar: Systematic Structure in a New Language," *Proceedings of the National Academy of Sciences* 102:2661–2665 (2005).
49. Nicholas Wade, "Deaf Children's Ad Hoc Language Evolves and Instructs," *New York Times*, September 21, 2004, p. F4.
50. Michael C. Corballis, "From Hand to Mouth," p. 205.
51. Louise Barrett, Robin Dunbar, and John Lycett, *Human Evolutionary Psychology*, Princeton University Press, 2002.
52. Geoffrey Miller, *The Mating Mind*, Random House, 2000.
53. Steven Pinker, "Language as an Adaptation to the Cognitive Niche," in Morten H. Christiansen and Simon Kirby, *Language Evolution*, Oxford University Press, 2003, p. 29.
54. Paul Mellars, "Neanderthals, Modern Humans and the Archaeological Evidence for Language," in Nina G. Jablonski and Leslie C. Aiello, eds., *The Origin and Diversification of Language*, California Academy of Sciences, 1988, p. 99.
55. Faraneh Vargha-Khadem, interview, October 1, 2001.
56. Faraneh Vargha-Khadem, Kate Watkins, Katie Alcock, Paul Fletcher, and Richard Passingham, "Praxic and Nonverbal Cognitive Deficits in a Large Family with a Genetically Transmitted Speech and Language Disorder," *Proceedings of the National Academy of Sciences* 92:930–933 (1995).
57. Faveneh Vargha-Khadem et al., "Neural Basis of an Inherited Speech and Language Disorder," *Proceedings of the National Academy of Science* 95:12659–12700 (1998).
58. Many genes were first discovered by biologists working with fruit flies, among whom it is a point of pride to give genes colorful names. These odd names are often adopted for the equivalent gene when it is discovered in humans, apart from names like *dunce* or *rutabaga* that are deemed too colorful. The forkhead gene was so named because when it is mutated, the fly larvae develop spiky structures on their head. Other genes of similar structure were found, and all turned out to have a signature section of DNA, called the forkhead box, which specified a region in the forkhead protein that binds to specific sequence of DNA. This is because the forkhead proteins control the activity of other genes by binding to regions of DNA just upstream of genes. Humans have turned out to possess a large family of forkhead box genes, the subgroups of which are named by letters of

the alphabet. FOXP2 is the second member of the P group of the family of forkhead box genes. Gary F. Marcus and Simon E. Fisher, "FOXP2 in Focus: What Can Genes Tell Us About Speech and Language?" *Trends in Cognitive Sciences* 7:257–262 (2003).

59. Cecilia S. L. Lai, Simon E. Fisher, Jane A. Hurst, Faraneh Vargha-Khadem, and Anthony P. Monaco, "A Forkhead-Domain Gene Is Mutated in a Severe Speech and Language Disorder," *Nature* 413:519–523 (2001).

60. Faraneh Vargha-Khadem et al., "FOXP2 and the Neuroanatomy of Speech and Language," *Nature Reviews Neuroscience* 6:131–138 (2005).

61. Wolfgang Enard et al., "Molecular Evolution of FOXP2, a Gene Involved in Speech and Language," *Nature* 418:869–872 (2002).

62. Svante Pääbo, interview, August 10, 2002.

63. Richard Klein, *The Human Career*, University of Chicago Press, 1999, p. 492.

64. The reason is probably that a rivalry between mitochondria inside the cell would be too disruptive. The sperm's mitochondria are made to carry a special chemical tag that says "kill me." As soon as the sperm enters the egg, its mitochondria are destroyed. The egg possesses about 100,000 mitochondria of its own, and has no need for the mere 100 or so contributed by the sperm. Douglas C. Wallace, Michael D. Brown, and Marie T. Lott, "Mitochondrial DNA Variation in Human Evolution and Disease," *Gene*, 238:211–230 (1999).

65. Peter A. Underhill et al., "Y Chromosome Sequence Variation and the History of Human Populations," *Nature Genetics*, 26:358–361 (2000).

66. S. T. Sherry, M. A. Batzer, and H. C. Harpending, "Modeling the Genetic Architecture of Modern Populations," *Annual Review of Anthropology* 27:153–169 (1968).

67. Jonathan K. Pritchard, Mark T. Seielstad, Anna Perez-Lezaun, and Marcus W. Feldman, "Population Growth of Human Y Chromosomes: A Study of Y Chromosome Micro-satellites," *Molecular Biology and Evolution* 16:1791–1798 (1999).

68. P. A. Underhill, G. Passarino, A. A. Lin, P. Shen, M. Mirazon-Lahr, R. A. Foley, P. J. Oefner, and L. L. Cavalli-Sforza, "The Phylogeography of Y Chromosome Binary Haplotypes and the Origins of Modern Human Populations," *Annals of Human Genetics* 65:43–62 (2001).

69. S. T. Sherry et al., "Modeling the Genetic Architecture of Modern Populations," p. 166.

70. Richard Borshay Lee, *The !Kung San*, Cambridge University Press, 1979, p. 31.

71. Tom Güldemann and Rainer Vossen, "Khoisan," in Bernd Heine and Derek Nurse, eds., *African Languages*, Cambridge University Press, 2000.

72. Yu-Sheng Chen et al., "mtDNA Variation in the South African Kung and Khwe—and Their Genetic Relationships to Other African Populations," *American Journal of Human Genetics* 66:1362–1383 (2000). The team sampled a group of !Kung from the northwestern Kalahari Desert known as the Vasikela !Kung.

73. Alec Knight et al., "African Y Chromosome and mtDNA Divergence Provides Insight into the History of Click Languages," *Current Biology* 13:464–473 (2003).

74. Nicholas Wade, "In Click Languages, an Echo of the Tongues of the Ancients," *New York Times*, March 18, 2003, p. F2.

75. Ornella Semino et al., "Ethiopians and Khoisan Share the Deepest Clades of the Human Y-Chromosome Phylogeny," *American Journal of Human Genetics* 70:265–268 (2002).

76. Donald E. Brown, *Human Universals*, McGraw-Hill, 1991, p. 139.

77. Henry Harpending and Alan R. Rogers, "Genetic Perspectives on Human Origins and Differentiation," *Annual Review of Genomics and Human Genetics* 1:361–385 (2000).

78. Richard Borshay Lee, *The !Kung San*, Cambridge University Press, 1979, p. 135.

79. Jon de la Harpe et al., "Diamphotoxin, the Arrow Poison of the !Kung Bushmen," *Journal of Biological Chemistry* 258:11924–11931 (1983).

80. Richard Borshay Lee, *The !Kung San*, p. 440.

81. Nancy Howell, *Demography of the Dobe !Kung,* Academic Press, 1979, p. 119.

82. "Harmless" is one interpretation of the Ju|'hoansi's name for themselves, as Elizabeth Marshall Thomas notes in a new book, *The Old Way.* Thomas argues that their way of life was "the most successful culture that our kind has ever known," as judged by its ecological stability and its endurance for at least 35,000 years. Her book's vivid personal account of the !Kung's hunter-gatherer lifestyle complements the anthropological study by Richard Borsay Lee. Elizabeth Marshall Thomas, *The Old Way — A Story of the First People,* Farrar Strauss Giroux, 2006.

83. Steven A. LeBlanc, *Constant Battles,* St. Martin's Press, 2003, p. 116.

84. Lawrence H. Keeley, *War before Civilization,* Oxford University Press, 1996, p. 134.

85. Richard Borshay Lee, *The !Kung San,* p. 399. The odd symbols represent different kinds of click.

86. Frank W. Marlowe, "Hunter-Gatherers and Human Evolution," *Evolutionary Anthropology* 14:54–67 (2005).

87. M. Siddall et al., "Sea-Level Fluctuations during the Last Glacial Cycle," *Nature* 423:853–858.

88. Exodus 15:8.

89. Sarah A. Tishkoff and Brian C. Verrelli, "Patterns of Human Genetic Diversity: Implications for Human Evolutionary History and Disease," *Annual Review of Genomics and Human Genetics* 4:293–340 (2003).

90. Jeffrey I. Rose, "The Question of Upper Pleistocene Connections between East Africa and South Arabia," *Current Anthropology* 45:551–555 (2004).

91. Luis Quintana-Murci et al., "Genetic Evidence of an Early Exit of Homo sapiens from Africa through Eastern Africa," *Nature Genetics* 23:437–441 (1999).

92. Martin Richards et al., "Extensive Female-Mediated Gene Flow from Sub-Saharan Africa into Near Eastern Arab Populations," *American Journal of Human Genetics* 72:1058–1064 (2003).

93. John D. H. Stead and Alec J. Jeffreys, "Structural Analysis of Insulin Minisatellite Alleles Reveals Unusually Large Differences in Diversity between Africans and Non-Africans," *American Journal of Human Genetics* 71:1273–1284 (2002).

94. Nicholas Wade, "To People the World, Start with 500," *New York Times,* November 11, 1997, p. F1.

95. Vincent Macaulay et al., "Single, Rapid Coastal Settlement of Asia Revealed by Analysis of Complete Mitochondrial Genomes," *Science* 308:1034–1036 (2005). The geneticists decoded the full mitochondrial DNA of people belonging to the lineages M and N, which are the earliest ones found outside Africa, and compared them with L3, the lineage inside Africa from which M and N are derived. Knowing the rate at which mutations occur in mitochondrial DNA, they calculated it would have taken 826 generations, or 20,650 years, for the M and N lineages to have evolved from L3. The time for this process to occur depends on the population size, and knowledge of the time allows the initial population to be estimated. The answer is 550 women of breeding age if the population remained the same size throughout the 20,000 years, and less than that if the population expanded, as was doubtless the case.

96. T. Kivisild et al., "The Genetic Heritage of the Earliest Settlers Persists Both in Indian Tribal and Caste Populations," *American Journal of Human Genetics* 72:313–332 (2003).

97. Richard G. Roberts et al., "New Ages for the Last Australian Megafauna: Continent-Wide Extinction About 46,000 Years Ago," *Science* 292:1888–1892 (2001).

98. Vincent Macaulay et al., "Single, Rapid Coastal Settlement of Asia."

99. Kirsi Huoponen, Theodore G. Schurr, Yu-Sheng Chen, and Douglas C. Wallace, "Mitochondrial DNA Variation in an Aboriginal Australian Population: Evidence for Genetic Isolation and Regional Differentiation," *Human Immunology* 62:954–969 (2001).

100. Max Ingman and Ulf Gyllensten, "Mitochondrial Genome Variation and Evolutionary History of Australian and New Guinean Aborigines," *Genome Research* 13:1600–1606 (2003).

101. Steven A. LeBlanc, *Constant Battles*, St. Martin's Press, 2003, p. 121.

102. Manfred Kayser et al., "Reduced Y-Chromosome, but Not Mitochondrial DNA, Diversity in Human Populations from West New Guinea," *American Journal of Human Genetics* 72:281–302 (2003).

103. Steven A. LeBlanc, *Constant Battles*, p. 151.

104. Nicholas Wade, "An Ancient Link to Africa Lives On in the Bay of Bengal," *New York Times*, December 10, 2002, p. A14.

105. Kumarasamy Thangaraj et al., "Genetic Affinities of the Andaman Islanders, a Vanishing Human Population," *Current Biology* 13:86–93 (2003).

106. Philip Endicott et al., "The Genetic Origins of the Andaman Islanders," *American Journal of Human Genetics* 72:1590–1593 (2003).

107. Madhusree Mukerjee, *The Land of Naked People*, Houghton Mifflin, 2003, p. 240.

108. Stephen Oppenheimer, *The Real Eve*, Carroll & Graf, 2003, p. 218.

109. David L. Reed et al., "Genetic Analysis of Lice Supports Direct Contact Between Modern and Archaic Humans," *Public Library of Science Biology* 2:1972–1983 (2004); Nicholas Wade, "What a Story Lice Can Tell," *New York Times*, October 5, 2004, p. F1. The second cluster of lice is found only in the Americas, suggesting that modern people contracted them from *Homo erectus* in East Asia or Siberia before crossing the Pleistocene-epoch land bridge that joined Siberia to North America. This lineage of lice may have been the dominant kind in the Americas until European colonists brought over the lineage standard in the rest of the world.

110. P. Brown et al., "A New Small-Bodied Hominin from the Late Pleistocene of Flores, Indonesia," *Nature* 431:1055–1091 (2004); Nicholas Wade, "New Species Revealed: Tiny Cousins of Humans," *New York Times*, October 28, 2004, p. A1; Nicholas Wade, "Miniature People Add Extra Pieces to Evolutionary Puzzle," *New York Times*, November 9, 2004, p. F2. M. J. Morwood et al., "Further evidence for small-bodied hominims from the Late Pleistocene of Flores, Indonesia," *Nature* 437, 1012–1017, 2005.

111. Argument about which species the little Floresians should be assigned to has continued to seethe, with several researchers suggesting the skull is a pathologically small modern human. A new position, developed by researchers at the Australian National University and the University of Sydney, is that the Floresians are neither a pathological version of *Homo sapiens* nor a downsized version of *Homo erectus*, but stem from an independent and much earlier migration out of Africa, perhaps before the island of Flores separated from the mainland. The argument is based on the skull's similarity in brain size and other features to *Homo ergaster*, the predecessor of *erectus*. Debbie Argue, Denise Donlon, Colin Groves, and Richard Wright, "*Homo floresiensis*: Microcephalic, pygmoid, *Australopithecus*, or *Homo*?," *Journal of Human Evolution* 51, 360–374 (2006).

112. Roger Lewin and Robert A. Foley, *Principles of Human Evolution*, Blackwell Science, 2004, p. 387.

113. Richard G. Klein, *The Human Career*, 2nd ed., University of Chicago Press, 1999, p. 470.

114. Christopher Stringer and Robin McKie, *African Exodus*, Henry Holt, 1996, 1999, p. 106.

115. Richard G. Klein, *The Human Career*, p. 477.

116. Matthias Krings et al., "Neanderthal DNA Sequences and the Origin of Modern Humans," *Cell* 90:19–30, 1997.

117. David Serre et al., "No Evidence of Neanderthal mtDNA Contribution to Early Modern Humans," *Public Library of Science Biology* 2:1–5 (2004).

118. A new theory of human origins proposes that the modern humans leaving Africa initially interbred with archaic humans as the emigrants' wave of advance engulfed the ar-

chaic populations. The theory predicts that a majority of sites on the nuclear genome may have some archaic alleles. Vinyarak Eswaran, Henry Harpending, and Alan R. Rogers, "Genomics Refutes an Exclusively African Origin of Humans," *Journal of Human Evolution* 49:1–18 (2005).

119. The roots of the Aurignacian culture are still obscure. If modern humans reached Europe from India, as seems likely from the genetic evidence, signs of predecessor culture might be expected in the Indian subcontinent. But so far the archaeological evidence from India shows little evidence of the sophisticated behaviors possessed by the Aurignacians. (Hannah V. A. James and Michael D. Petraglia, "Modern Human Origins and the Evolution of Behavior in the Later Pleistocene Record of South Asia," *Current Anthropology*, 46 [Supplement]: 3–27 [2005]). Possibly the elements of the Aurignacian culture were formed at some point during the migration from India to Europe, as the first modern humans adapted from a subtropical climate to that of the European ice age.

120. The Aurignacian tools at this site now seem to have been made by modern humans who occupied the site in between periods of Neanderthal occupation. Brad Gravina, Paul Mellars and Christopher Bronk Ramsey, "Radiocarbon Dating of Interstratified Neanderthal and Early Modern Human Occupations at the Chatelperronian Type-Site," *Nature* 438:51–56 (2005).

121. Martin Richards, "The Neolithic Invasion of Europe," *Annual Review of Anthropology* 32:135–162 (2003).

122. An important revision in radiocarbon dating indicates an earlier and much compressed time scale for the Aurignacians' spread across Europe. The revision was prompted by a) a new method of filtering out contaminants that have made ancient carbon sources seem younger than they were, and b) a new estimate of the amount of carbon-14 in the atmosphere in the distant past. With these two refinements, Paul Mellars has now revised his timetable for the arrival of modern humans in Europe. He estimates that they had arrived west of the Black Sea by 46,000 years ago, not 40,000–44,000 years ago as shown in Figure 5.2, and had reached northern Spain by 41,000 years ago, not 36,000 years ago. Mellars argues that the Neanderthal fossils from the Zafarraya cave in Spain and Vindija in Croatia, both at present dated to about 30,000 years ago, will turn out to be much older. If so, the new timetable indicates that the Neanderthals succumbed much more quickly than had been supposed, perhaps in a mere 5,000 years. The moderns' rapid advance may have been helped by an improvement in climatic conditions that occurred between 43,000 and 41,000 years ago. Paul Mellars, "A New Radiocarbon Revolution and the Dispersal of Modern Humans in Eurasia," *Nature* 439:931–935 (2006).

123. Ofer Bar-Yosef, "The Upper Paleolithic Revolution," *Annual Review of Anthropology* 31:363–393 (2002).

124. Agnar Helgason et al., "An Icelandic Example of the Impact of Population Structure on Association Studies," *Nature Genetics* 37:90–95, 2005; Nicholas Wade, "Where Are You From? For Icelanders, the Answer Is in the Genes," *New York Times*, December 28, 2004, p. F3.

125. Patrick D. Evans et al., "Microcephalin, a Gene Regulating Brain Size, Continues to Evolve Adaptively in Humans," *Science* 309:1717–1220 (2005); Nitzan Mekel-Bobrov, "Ongoing Adaptive Evolution of ASPM, a Brain Size Determinant in Homo sapiens," *Science* 309:1720–1722 (2005).

126. Nicholas Wade, "Brain May Still Be Evolving, Studies Hint," *New York Times*, September 9, 2005, p. A14.

127. Steve Dorus et al., "Accelerated Evolution of Nervous System Genes in the Origin of Homo sapiens," *Cell* 119:1027–1040 (2004).

128. Lahn's two microcephalin genes did not show up in a genomewide search for recently selected genes performed by Jonathan Pritchard and colleagues (see note 256). That may

reflect limitations of the Pritchard test, no test for selection being perfect. Pritchard did detect signals of selection in two other genes involved in microcephaly, and in several other types of brain gene. Some of these brain genes were under selection in Africans, some in Asians, and some in Europeans, confirming Lahn's view that cognitive evolution may have proceeded independently in the three populations. Many of the genetic changes occurring independently in the major continental races are likely to have been convergent, meaning that evolution was using the different mutations available to it in each population to bring about the same adaptation.

129. A new refinement of the radiocarbon dating method (see note 122 above) indicates the drawings of the Chauvet cave are much older than supposed. The first occupation can now be dated to 36,000 years ago.

130. Richard G. Klein, *The Human Career*, 2nd edition, University of Chicago Press, 1999, p. 540.

131. Martin Richards et al., "Tracing European Founder Lineages in the Near Eastern mtDNA Pool," *American Journal of Human Genetics* 67:1251–1276 (2000); Martin Richards, "The Neolithic Invasion of Europe," *Annual Review of Anthropology* 32:135–162 (2003).

132. Antonio Torroni et al., "A Signal, from Human mtDNA, of Postglacial Recolonization in Europe," *American Journal of Human Genetics* 69:844–852 (2001).

133. Ornella Semino et al., "The Genetic Legacy of Paleolithic *Homo sapiens sapiens* in Extant Europeans: A Y Chromosome Perspective," *Science* 290:1155–1159, 2000.

134. Colin Renfrew, "Commodification and Institution in Group-Oriented and Individualizing Societies," in *The Origin of Human Social Institutions*, Oxford University Press, 2001, p. 114.

135. Carles Vilà et al., "Multiple and Ancient Origins of the Domestic Dog," *Science* 276:1687–1689 (1997).

136. Peter Savolainen et al., "Genetic Evidence for an East Asian Origin of Domestic Dogs," *Science* 298:1610–1616 (2002).

137. Lyudmila N. Trut, Early Canid Domestication: The Farm-Fox Experiment," *American Scientist* 17:160–169 (1999).

138. Brian Hare, Michelle Brown, Christine Williamson, and Michael Tomasello, "The Domestication of Social Cognition in Dogs," *Science* 298:1634–1636 (2002).

139. Nicholas Wade, "From Wolf to Dog, Yes, but When?" *New York Times*, November 22, 2002, p. A20.

140. Jennifer A. Leonard et al., "Ancient DNA Evidence for Old World Origin of New World Dogs," *Science* 298:1613–1616 (2002).

141. Joseph H. Greenberg, *Language in the Americas*, Stanford University Press, 1987, p. 43.

142. Vivian Scheinsohn, "Hunter-Gatherer Archaeology in South America," *Annual Review of Anthropology* 32:339–361 (2003).

143. Yelena B. Starikovskaya et al., "mtDNA Diversity in Chukchi and Siberian Eskimos: Implications for the Genetic History of Ancient Beringia and the Peopling of the New World," *American Journal of Human Genetics* 63:1473–1491 (1998).

144. Mark Seielstad et al., "A Novel Y-Chromosome Variant Puts an Upper Limit on the Timing of First Entry into the Americas," *American Journal of Human Genetics* 73:700–705 (2003).

145. Maria-Catira Bortolini et al., "Y-Chromosome Evidence for Differing Ancient Demographic Histories in the Americas," *American Journal of Human Genetics* 73:524–539 (2003).

146. Michael D. Brown et al., "mtDNA Haplogroup X: An Ancient Link between Europe/ Western Asia and North America?" *American Journal of Human Genetics* 63:1852–1861 (1998).

147. Ripan S. Malhi and David Glenn Smith, "Haplogroup X Confirmed in Prehistoric North America," *American Journal of Physical Anthropology* 119:84–86 (2002).

148. Eduardo Ruiz-Pesini et al., "Effects of Purifying and Adaptive Selection on Regional Variation in Human mtDNA," *Science* 303:223–226; Dan Mishmar et al., "Natural Selection Shaped Regional mtDNA Variation in Humans." *Proceedings of the National Academy of Sciences* 100:171–176 (2003).

149. Mark T. Seielstad, Erich Minch, and L. Luca Cavalli-Sforza, "Genetic Evidence for a Higher Female Migration Rate in Humans," *Nature Genetics* 20:278–280 (1998).

150. Kennewick Man, the 9,000–year-old, non-Mongoloid-looking skeleton discovered in Washington state and claimed by sinodont American Indians as their intimate ancestor, is a sundadont.

151. Marta Mirazón Lahr, *The Evolution of Modern Human Diversity*, Cambridge University Press, 1996, p. 318.

152. Increasing evidence suggests that pale skin, a variation on the dark skin of the ancestral human population, arose independently in the populations of west and east Eurasia, even though the cause—adapting to the reduced sunlight of northern latitudes—was presumably the same in both cases. Five genes affecting skin color show signs of recent selection in Europeans but not in East Asians. (See notes 241 and 256.) This implies either that East Asians acquired their pale skin through changes in a different set of genes or that their skins became pale considerably earlier than did those of Europeans. In the latter case, the signs of selection would have faded and the genes in East Asians would not have been flagged by Pritchard's test.

153. Most people in Africa and Europe have wet earwax. But dry earwax is the rule among East Asians. A team of Japanese researchers has traced the difference to a mutation in a gene called ABCC11. (Koh-ichiro Yoshiura et al., "A SNP in the ABCC11 Gene Is the Determinant of Human Earwax Type," *Nature Genetics* 38:324–330 [2006].) The mutation seems to have arisen in the northern part of east Asia and to have become common very quickly—it is almost universal in northern Han Chinese and in Koreans. What selection pressure made the new version of the gene spread so rapidly? Earwax serves the very humble role of biological flypaper—it stops insects and dirt from getting into the ear—and a change in earwax consistency seems unlikely to have been much of an advantage. The new gene was probably selected because it seems also to reduce the amount of sweating, and hence of body odor. One or the other quality, or both, may have given the gene its decisive advantage.

154. Richard G. Klein, *The Human Career*, p. 502.

155. Ofer Bar-Yosef, "On the Nature of Transitions: The Middle to Upper Paleolithic and the Neolithic Revolution," *Cambridge Journal of Archaeology* 8:(2):141–163 (1998).

156. Ofer Bar-Yosef, "From Sedentary Foragers to Village Hierarchies," *The Origin of Human Social Institutions*, Oxford University Press, 2001, p. 7.

157. Peter M. M. G. Akkermans and Glenn M. Schwartz, *The Archaeology of Syria*, Cambridge University Press, 2003, p. 45.

158. Brian M. Fagan, *People of the Earth*, 10th ed., Prentice Hall, 2001, p. 226.

159. The evidence included a Natufian skeleton with a Natufian type arrow point embedded in the thoracic vertebrae. Fanny Bocquentin and Ofer Bar-Yosef, "Early Natufian Remains: Evidence for Physical Conflict from Mt. Carmel, Israel," *Journal of Human Evolution* 47:19–23 (2004).

160. Akkermans and Schwartz, *The Archaeology of Syria*, p. 96.

161. Colin Renfrew, "Commodification and Institution in Group-Oriented and Individualizing Societies," in W. G. Runciman, ed., *The Origin of Social Institutions*, Oxford University Press, 2001, p. 95.

162. Natalie D. Munro, "Zooarchaeological Measures of Hunting Pressure and Occupation Intensity in the Natufian," *Current Anthropology* 45:S5–33 (2004).

163. Akkermans and Schwartz, *The Archaeology of Syria*, p. 70.

164. Daniel Zohary and Maria Hopf, *Domestication of Plants in the Old World*, 3rd ed., Oxford University Press, 2000, p. 18.

165. Robin Allaby, "Wheat Domestication," in *Archaeogenetics: DNA and the Population Prehistory of Europe*, McDonald Institute for Archaeological Research, 2000, pp. 321–324.

166. Francesco Salamini et al., "Genetics and Geography of Wild Cereal Domestication in the Near East," *Nature Reviews Genetics* 3:429–441 (2002).

167. Manfred Heun et al., "Site of Einkorn Wheat Domestication Identified by DNA Fingerprinting," *Science* 278:1312–1314 (1997); Daniel Zohary, and Maria Hopf, *Domestication of Plants in the Old World*, p. 36.

168. Francesco Salamini et al., "Genetics and Geography of Wild Cereal Domestication in the Near East."

169. Christopher S. Troy et al., "Genetic Evidence for Near-Eastern Origins of European Cattle," *Nature* 410:1088–1091 (2001).

170. Carlos Vilà et al., "Widespread Origins of Domestic Horse Lineages," *Science* 291:474–477 (2001).

171. However, the horse Y chromosome is telling a possibly different story. Unlike the mitochondrial DNA, samples of a small region of the Y chromosome from 15 different breeds of European and Asian horses proved to be identical. Gabriella Lindgren et al., "Limited Number of Patrilines in Horse Domestication," *Nature Genetics* 36:335–336 (2004). This indicates that ancient horse breeders, like modern ones, often let a single stallion cover many females. It could also allow for the possibility that independent domestications of the horse were not as common as the mitochondrial DNA evidence suggests.

172. This assumption has been confirmed by a direct study of the first Neolithic farmers to settle in Europe. Mitochondrial DNA was extracted from bones taken from archaeological sites of the earliest farming communities of 7,500 years ago in Germany, Austria, and Hungary. A quarter of the sample belonged to the N1a subbranch of the mitochondrial tree, a type that is now very rare among Europeans. This implies that though Neolithic farming techniques spread rapidly, the farmers themselves did not make much of a genetic impact. The authors favor the possibility "that small pioneer groups carried farming into new areas of Europe, and that once the technique had taken root, the surrounding hunter-gatherers adopted the new culture and then outnumbered the original farmers, diluting their N1a frequency to the low modern value." Wolfgang Haak et al., "Ancient DNA from the First European Farmers in 7500-Year-Old Neolithic Sites," *Science* 310:1016–1018 (2005).

173. Ornella Semino et al., "The Genetic Legacy of Paleolithic *Homo sapiens sapiens* in Extant Europeans: A Y Chromosome Perspective," *Science* 290:1155–1159 (2000).

174. Roy King and Peter A. Underhill, "Congruent Distribution of Neolithic Painted Pottery and Ceramic Figurines with Y-Chromosome Lineages," *Antiquity* 76:707–714 (2002).

175. I thank a reader, the food writer Anne Mendelson, for pointing out that in many parts of the world milk is consumed only in sour form, such as yogurt, after its lactose has been converted by bacteria into lactic acid. In such conditions there would be no pressure for lactose tolerance to evolve. Nonetheless, lactose tolerance did arise, so presumably the people of the Funnel Beaker culture must have consumed milk in raw form, perhaps not knowing how to ferment it.

176. Albano Beja-Pereira et al., "Gene-Culture Coevolution between Cattle Milk Protein Genes and Human Lactase Genes," *Nature Genetics* 35:311–315 (2003).

177. Nabil Sabri Enattah et al., "Identification of a Variant Associated with Adult-type Hypolactasia," *Nature Genetics* 30:233–237 (2002).

178. Todd Bersaglieri et al., "Genetic Signatures of Strong Recent Positive Selection at the Lactase Gene," *American Journal of Human Genetics* 74:1111–1120 (2004).

179. Charlotte A. Mulcare et al., "The T Allele of a Single-Nucleotide Polymorphism 13.9 kb Upstream of the Lactase Gene (LCT) (C-13.9kbT) Does Not Predict or Cause the Lactase-Persistence Phenotype in Africans," *American Journal of Human Genetics* 74:1102–1110 (2004).

180. Napoleon Chagnon, "Life Histories, Blood Revenge, and Warfare in a Tribal Population," *Science* 239:985–992 (1988).

181. Anne E. Pusey, in Frans B. M. de Waal, ed., *Tree of Origin*, Harvard University Press 2001, p. 21.

182. John C. Mitani, David P. Watts, and Martin N. Muller, "Recent Developments in the Study of Wild Chimpanzee Behavior," *Evolutionary Anthropology* 11: 9–25 (2002).

183. Anne E. Pusey, *Tree of Origin*, p. 26.

184. Julie L. Constable, Mary V. Ashley, Jane Goodall, and Anne E. Pusey, "Noninvasive Paternity Assignment in Gombe Chimpanzees," *Molecular Ecology* 10:1279–1300 (2001).

185. Anne Pusey, Jennifer Williams, and Jane Goodall, "The Influence of Dominance Rank on the Reproductive Success of Female Chimpanzees," *Science* 277:828–831 (1997).

186. A. Whiten et al., "Cultures in Chimpanzees," *Nature* 399:682 (1999).

187. W. C. McGrew, "Tools Compared," in Richard W. Wrangham et al., eds., *Chimpanzee Cultures*, Harvard University Press, 1994, p. 25.

188. Richard W. Wrangham et al., *Demonic Males*, Houghton Mifflin, 1996, p. 226.

189. Some experts argue that chimps are derived from bonobos, but the direction of change makes no difference here.

190. Frans de Waal, *Our Inner Ape*, Riverhead Books, 2005, p. 221.

191. John C. Mitani, David P. Watts, and Martin N. Muller, "Recent Developments in the Study of Wild Chimpanzee Behavior," *Evolutionary Anthropology* 11:9–25 (2002).

192. Napoleon A. Chagnon, *Yanomamo*, 5th ed., Wadsworth, 1997, p. 189.

193. Ibid., p. 97.

194. Lawrence H. Keeley, *War before Civilization*, Oxford University Press, 1996, p. 174.

195. Ibid., p. 33.

196. Steven A. LeBlanc, *Constant Battles*, St. Martin's Press, 2003, p. 8.

197. Simon Mead et al., "Balancing Selection at the Prion Protein Gene Consistent with Prehistoric Kurulike Epidemics," *Science* 300:640–643 (2003).

198. Edward O. Wilson, *On Human Nature*, Harvard University Press, 1978, p. 114.

199. Louise Barrett, Robin Dunbar, and John Lycett, *Human Evolutionary Psychology*, Princeton University Press, 2002, p. 64.

200. Napoleon A. Chagnon, *Yanomamo*, 5th ed., Wadsworth, 1997, p. 76.

201. Napoleon Chagnon, "Life Histories, Blood Revenge, and Warfare in a Tribal Population," *Science* 239:985–992 (1988).

202. Napoleon A. Chagnon, *Yanomamo*, p. 77.

203. David M. Buss, *Evolutionary Psychology*, 2nd ed., Pearson Education, 2004, p. 257.

204. Robert L. Trivers, "The Evolution of Reciprocal Altruism," *Quarterly Review of Biology* 46:35–57, 1971.

205. Following Steven Pinker, *How the Mind Works*, W. W. Norton, 1997, pp. 404–405.

206. Matt Ridley, *The Origins of Virtue*, Viking, 1996, p. 197.

207. Paul Seabright, "The Company of Strangers: A Natural History of Economic Life," Princeton University Press, 2004, p. 28.

208. Michael Kosfeld et al., "Oxytocin Increases Trust in Humans," *Nature* 435:673–676 (2005).

209. Roy A. Rappaport, "The Sacred in Human Evolution," *Annual Review of Ecology and Systematics* 2:23–44 (1971).

210. Joyce Marcus and Kent V. Flannery, "The Coevolution of Ritual and Society: New ^{14}C Dates from Ancient Mexico," *Proceedings of the National Academy of Sciences* 18252–18261 (2004).

211. Richard Sosis, "Why Aren't We all Hutterites?" *Human Nature* 14:91–127 (2003).

212. Edward O. Wilson, *On Human Nature*, Harvard University Press, 1978, p. 175.

213. Frank W. Marlowe, "A Critical Period for Provisioning by Hadza Men: Implications for Pair Bonding," *Evolution and Human Behavior* 24:217–229 (2003).

214. Tim Birkhead, *Promiscuity*, Harvard University Press, 2000, p. 41.

215. Nicholas Wade, "Battle of the Sexes is Discerned in Sperm," *New York Times*, February 22, 2000, p. F1.

216. Alan F. Dixson, *Primate Sexuality*, Oxford University Press, 1998, p. 218.

217. Gerald J. Wyckoff, Wen Wang, and Chung-I Wu, "Rapid Evolution of Male Reproductive Genes in the Descent of Man," *Nature* 403:304–309 (2000).

218. Sperm generally survive for less than forty-eight hours in the human female reproductive tract, although survival times up to five days occasionally occur. Tim Birkhead, *Promiscuity*, p. 67.

219. Robin Baker, *Sperm Wars*, Basic Books, 1996, p. 38. Baker has contributed several novel findings to this field but some have not survived challenge by other researchers; see Birkhead, as cited, pp. 23–29.

220. W. H. James, "The Incidence of Superfecundation and of Double Paternity in the General Population," *Acta Geneticae Medicae et Gemellologiae* 42:257–262 (1993).

221. R. E. Wenk et al., "How Frequent is Heteropaternal Superfecundation?" *Acta Geneticae Medicae et Gemellologiae* 41:43–47 (1992).

222. Louise Barrett, Robin Dunbar, and John Lycett, *Human Evolutionary Psychology*, Princeton University Press, 2002, p. 181.

223. Bobbi S. Low, *Why Sex Matters*, Princeton University Press, 2000, p. 80.

224. David M. Buss, *Evolutionary Psychology*, 2nd ed., Pearson Education, 2004, p. 112.

225. Ibid., p. 117.

226. Geoffrey Miller, *The Mating Mind: How Sexual Choice Shaped the Evolution of Human Nature*, Doubleday, 2000.

227. Marta Mirazón Lahr, *The Evolution of Modern Human Diversity: A Study of Cranial Variation*, Cambridge University Press, 1996, p. 263.

228. Ibid., p. 337.

229. Marta Mirazón Lahr and Richard V. S. Wright, "The Question of Robusticity and the Relationship Between Cranial Size and Shape in *Homo sapiens*," *Journal of Human Evolution* 31:157–191 (1996).

230. Helen M. Leach, "Human Domestication Reconsidered," *Current Anthropology* 44:349–368 (2003).

231. Ibid., p. 360.

232. Richard Wrangham, interview, *Edge*, February 2, 2002, www.edge.org.

233. Allen W. Johnson and Timothy Earle, *The Evolution of Human Societies*, 2nd ed., Stanford University Press, 2000.

234. www.who.int/whr/2004/annex/topic/en/annex_2_en.pdf

235. Derek V. Exner et al., "Lesser Response to Angiotensin-converting-enzyme Inhibitor Therapy in Black as Compared with White Patients with Left Ventricular Dysfunction, *New England Journal of Medicine* 344:1351–1357 (2001).

236. Anne L. Taylor et al., "Combination of Isosorbide Dinitrate and Hydralazine in Blacks with Heart Failure," *New England Journal of Medicine* 351:2049–2057, 2004; Nicholas Wade, "Race-Based Medicine Continued," *New York Times* November 14, 2004, Section 4, p. 12.

237. Robert S. Schwartz, "Racial Profiling in Medical Research," *New England Journal of Medicine* 344:1392–1393 (2001).

238. "Genes, Drugs and Race," *Nature Genetics* 29:239 (2001).

239. Neil Risch, Esteban Burchard, Elav Ziv, and Hua Tang, "Categorization of Humans in Biomedical Research: Genes, Race and Disease," genomebiology.com/2002/3/7/comment/2007

240. Agnar Helgason et al., "An Icelandic Example of the Impact of Population Structure on Association Studies," *Nature Genetics* 37:90–95 (2005); Nicholas Wade, "Where Are You From? For Icelanders, the Answer Is in the Genes," *New York Times*, December 28, 2004, p. F3.

241. Rebecca L. Lamason et al., "SLC24A5, a Putative Cation Exchanger, Affects Pigmentation in Zebrafish and Humans," *Science* 310:1782–1786 (2005).

242. Noah A. Rosenberg et al., "Genetic Structure of Human Populations," *Science* 298:2381–2385 (2002).

243. Nicholas Wade, "Gene Study Identifies 5 Main Human Populations, Linking Them to Geography," *New York Times*, December 20, 2002, p. A37.

244. In the forensic system used in the United States, a suspect's genome is analyzed at 13 specific sites and the number of repeats is counted. At any one site, there may be many people who possess the same number of repeats. Far fewer people have the same number at two sites. And the likelihood of any two people in the U.S. population having the same repeat number at all 13 sites is so small that everyone save identical twins is assumed to have their own unique set of repeats. (The number of repeats at each site is in fact a pair of numbers, one from the chromosome inherited from the mother, the other from the father's, but they are often the same.)

245. Nicholas Wade, "For Sale: A DNA Test to Measure Racial Mix," *New York Times*, October 1, 2002, p. F4.

246. Nicholas Wade, "Unusual Use of DNA Aided in Serial Killer Search," *New York Times*, June 2, 2003, p. A28.

247. Richard Lewontin, *Human Diversity*, W. H. Freeman, 1995, p. 123.

248. Statement adopted by the council of the American Sociological Association, August 9, 2002; available on www.asanet.org.

249. American Anthropological Association statement on "Race," May 17, 1998; www.aaanet.org.

250. Quoted by Henry Harpending and Alan R. Rogers in "Genetic Perspectives on Human Origins and Differentiation," *Annual Review of Genomics and Human Genetics* 1:361–385 (2000).

251. David L. Hartl and Andrew G. Clark, *Principles of Population Genetics*, 3rd ed., Sinauer Associates, 1997, p. 119.

252. Patrick D. Evans et al., "Microcephalin, a Gene Regulating Brain Size, Continues to Evolve Adaptively in Humans," *Science* 309:1717–1720 (2005).

253. Nitzan Mekel-Bobrov, "Ongoing Adaptive Evolution of ASPM, a Brain Size Determinant in *Homo sapiens*," *Science* 309:1720–1722 (2005).

254. John H. Relethford, "Apportionment of Global Human Genetic Diversity Based on Craniometrics and Skin Color," *American Journal of Physical Anthropology* 118:393–398 (2002).

255. Henry Harpending and Alan R. Rogers, "Genetic Perspectives on Human Origins and Differentiation," 1:380.

256. The test by Pritchard and his colleagues was based on the fact that a beneficial mutation is inherited in a large block of DNA, which will carry its own signature set of DNA changes. If the gene with the good mutation spreads rapidly, along with its block, the DNA in that region of the chromosome will become less diverse in the population as a whole because so many people now carry the same sequence of DNA units at that location.

Pritchard's test measures the difference in diversity between those who carry a new version of a gene and those who do not. Lesser diversity in a population is taken as a sign of selection. The difference in diversity disappears as the gene becomes universal, because increasing numbers of people carry the new variant. Thus the test picks up only new gene variants on their way to becoming universal, i.e., recently selected genes.

Pritchard looked for blocks with selected genes in data gathered by the Hap Map project from Africans (specifically, the Yoruba people of Nigeria), East Asians, and Europeans. The time of the selective pressure is about 10,800 years ago for the African genes and 6,600 for the East Asian and European genes.

As shown in the figure, 206 selected genetic regions were identified in Africans, 185 in East Asians, and 188 in Europeans. The fact that the selected genes do not overlap very much indicates that each population evolved independently. The selected genes shared by two races may have arisen by migration or be instances of independent evolution.

The genes under selection clustered in specific categories. Some of the strongest signals of selection were for 4 genes for skin color, found to be under selection in Europeans but not in Asians. Another category of selected genes was in those for skeletal development (possibly reflecting the gracilization of human populations). Other selec-

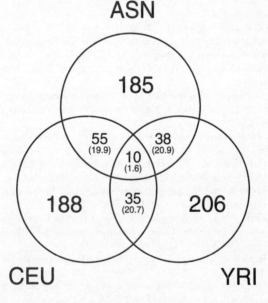

FIGURE 9.1

Genes that have undergone recent evolutionary change in the genomes of East Asians (ASN), Europeans (CEU), and Africans (YRI).

tion categories were genes for fertility, for taste and smell, and genes involved in the metabolism of foodstuffs. The latter two groups may reflect the sharp changes in diet that followed the Neolithic revolution. Benjamin Voight, Sridhar Kudaravalli, Xiaoquan Wen, and Jonathan K. Pritchard, "A Map of Recent Positive Selection in the Human Genome," *PloS Biology* 4:446–458 (2006).

257. Nicholas Wade, "Race-Based Medicine Continued," *New York Times*, November 14, 2004, Section 4, p. 12.

258. Jon Entine, *Taboo*, Public Affairs, 2000, p. 34.

259. Ibid., pp. 39–40.

260. Jon Entine, "The Straw Man of 'Race,'" *World & I*, September 2001, p. 309.

261. John Manners, "Kenya's Running Tribe," available online at www.umist.ac.uk

262. Ibid.

263. Jared Diamond, *Guns, Germs, and Steel*, W. W. Norton, 1997, p. 25.

264. Richard Klein, *The Human Career*, 2nd ed., University of Chicago Press, 1999, p. 502.

265. Robin I. M. Dunbar, "The Origin and Subsequent Evolution of Language," in Morten H. Christiansen and Simon Kirby, *Language Evolution*, Oxford University Press, 2003, p. 231.

266. Judges 12:5–6.

267. Denis Mack Smith, *Medieval Sicily*, Chatto & Windus, 1968, p. 71.

268. William A. Foley, "The Languages of New Guinea," *Annual Review of Anthropology* 29:357–404 (2000).

269. Jonathan Adams and Marcel Otte, "Did Indo-European Languages Spread before Farming?" *Current Anthropology* 40:73–77 (1999).

270. Jared Diamond and Peter Bellwood, "Farmers and Their Languages: The First Expansions," *Science* 300:597–603 (2003).

271. Christopher Ehret, O. Y. Keita, and Paul Newman, "Origins of Afroasiatic," *Science* 306:1680–1681 (2004).

272. Christopher Ehret, "Language Family Expansions: Broadening Our Understandings of Cause from an African Perspective," in Peter Bellwood and Colin Renfrew, eds., *Examining the Farming/Language Dispersal Hypothesis*, McDonald Institute for Archaeological Research, 2002, p. 173.

273. Colin Renfrew, *Archaeology and Language*, Jonathan Cape, 1987.

274. L. Luca Cavalli-Sforza, Paolo Menozzi, and Alberto Piazza, *The History and Geography of Human Genes*, Princeton University Press, 1994, pp. 296–299.

275. Ibid., p. 299.

276. Marek Zvelebil, "Demography and Dispersal of Early Farming Populations at the Mesolithic-Neolithic Transition: Linguistic and Genetic Implications," in Peter Bellwood and Colin Renfrew, ed., *Examining the Farming/Language Dispersal Hypothesis*, McDonald Institute for Archaeological Research, 2002, p. 381.

277. Colin Renfrew, April McMahon, and Larry Trask, eds., *Time Depth in Historical Linguistics*, McDonald Institute for Archaeological Research, 2000.

278. Bill J. Darden, "On the Question of the Anatolian Origin of Indo-Hittite," in *Great Anatolia and the Indo-Hittite Language Family*, Journal of Indo-European Studies Monograph No. 38, Institute for the Study of Man, 2001.

279. Mark Pagel, "Maximum-Likelihood Methods for Glottochronology and for Reconstructing Linguistic Phylogenies," in Colin Renfrew, April McMahon, and Larry Trask, eds., *Time Depth in Historical Linguistics*, McDonald Institute for Archaeological Research, 2000, p. 198.

280. K. Bergsland and H. Vogt, "On the Validity of Glottochronology," *Current Anthropology* 3:115–153 (1962).

281. Russell D. Gray and Quentin D. Atkinson, "Language-Tree Divergence Times Support the Anatolian Theory of Indo-European Origin," *Nature* 426:435–439 (2003).

282. Peter Forster and Alfred Toth, "Toward a Phylogenetic Chronology of Ancient Gaulish, Celtic and Indo European," *Proceedings of the National Academy of Sciences* 100:9079–9084 (2003).

283. Bernd Heine and Derek Nurse, eds., *African Languages*, Cambridge University Press, 2000, p. 1.

284. Christopher Ehret, "Language and History," in Bernd Heine and Derek Nurse, eds., *African Languages*, Cambridge University Press, 2000, pp. 272–298; "Testing the Expectations of Glottochronology against the Correlations of Language and Archaeology in Africa," in Colin Renfrew, April McMahon, and Larry Trask, eds., *Time Depth in Historical Linguistics*, McDonald Institute for Archaeological Research, 2000, pp. 373–401.

285. Richard J. Hayward, "Afroasiatic," in Bernd Heine and Derek Nurse, eds., *African Languages*, Cambridge University Press, 2000, pp. 79–98.

286. Christopher Ehret, "Language and History," in Bernd Heine and Derek Nurse, eds., *African Languages*, Cambridge University Press, 2000, p. 290.

287. Nicholas Wade, "Joseph Greenberg, 85, Singular Linguist, Dies," *New York Times*, May 15, 2001, p. A23.

288. Joseph H. Greenberg, "Language in the Americas," Stanford University Press, 1987.

289. L. Luca Cavalli-Sforza, Paolo Menozzi, and Alberto Piazza, *The History and Geography of Human Genes*, Princeton University Press, 1994, pp. 317, 96.

290. Ibid., p. 99.

291. L. L. Cavalli-Sforza, Eric Minch, and J. L. Mountain, "Coevolution of Genes and Languages Revisited," *Proceedings of the National Academy of Sciences* 89:5620–5624, (1992).

292. M. Lionel Bender, in Bernd Heine and Derek Nurse, eds., *African Languages*, Cambridge University Press, 2000, p. 54.

293. The example is adapted from Lyle Campbell, *Historical Linguistics*, MIT Press, 1999, p. 111.

294. Richard Hayward, in Bernd Heine and Derek Nurse, eds., *African Languages*, Cambridge University Press, 2000, p. 86.

295. Nicholas Wade, "Scientist at Work: Joseph H. Greenberg; What We All Spoke When the World Was Young," *New York Times*, February 1, 2000, p. F1.

296. Joseph H. Greenberg, *Indo-European and Its Closest Relatives, Volume 1: Grammar*, Stanford University Press, 2000, p. 217.

297. Joseph H. Greenberg, *Indo-European and Its Closest Relatives, Volume 2: Lexicon*, Stanford University Press, 2002.

298. Ibid., p. 2.

299. Nicholas Wade, "Joseph Greenberg, 85, Singular Linguist, Dies," *New York Times*, May 15, 2001, p. A23; Harold C. Fleming, "Joseph Harold Greenberg: A Tribute and an Appraisal," *Mother Tongue: The Journal* VI:9–27 (2000–2001).

300. Johanna Nichols, "Modeling Ancient Population Structures and Movement in Linguistics," *Annual Review of Anthropology*, 26:359–384 (1979).

301. Mark Pagel, "Maximum-Likelihood Methods for Glottochronology and for Reconstructing Linguistic Phylogenies."

302. Merritt Ruhlen, *The Origin of Language*, Wiley, 1994, p. 115.

303. Tatiana Zerjal et al., "The Genetic Legacy of the Mongols," *American Journal of Human Genetics* 72:717–721 (2003).

304. Ala-ad-Din 'Ata-Malik Juvaini, *The History of the World Conqueror*, trans. J. A. Boyle, Manchester University Press, 1958, vol. 2, p. 594.

305. Nicholas Wade, "A Prolific Genghis Khan, It Seems, Helped People the World," *New York Times*, February 11, 2003, p. F3.

306. Yali Xue et al., "Recent Spread of a Y-Chromosomal Lineage in Northern China and Mongolia," *American Journal of Human Genetics*, 77:1112–1116 (2005).

307. A remarkable 20% of men in northwestern Ireland carry a particular set of mutations on their Y chromosome, known as the Irish modal haplotype. Many have surnames that are associated with the Ui Neill, a group of dynasties that claimed the high kingship of Ireland and ruled the northwest and other parts of Ireland from about AD 600 to 900. Ui Neill means "descendants of Niall." Historians have tended to regard the Ui Neill as a political construct and its patriarch, Niall of the Nine Hostages, as a probably legendary figure. The genetic evidence provides striking evidence that Niall really existed, a finding as surprising as if the legend of King Arthur turned out to be solid history. The Irish modal haplotype, the signature of descent from Niall, is most common in northwestern Ireland but is also found in the Irish diaspora, being carried by no less than 2% of New Yorkers of European descent. Evidently one should listen less skeptically to Irishmen who declare the blood of Irish kings runs in their veins. Laoise T. Moore et al., "A Y-Chromosome Signature of Hegemony in Gaelic Ireland," *American Journal of Human Genetics* 78:334–338 (2006).

308. George Redmonds, interview, April 5, 2000.

309. Bryan Sykes, *Adam's Curse*, W. W. Norton, 2004, p. 7.

310. Bryan Sykes and Catherine Irven, "Surnames and the Y Chromosome," *American Journal of Human Genetics* 66:1417–1419 (2000).

311. Nicholas Wade, "If Biology Is Ancestry, Are These People Related?" *New York Times*, April 9, 2000, Section 4, p. 4.

312. Bryan Sykes, *Adam's Curse*, p. 18.

313. Cristian Capelli et al., "A Y Chromosome Census of the British Isles," *Current Biology* 13:979–984 (2003).

314. Emmeline W. Hill, Mark A. Jobling, and Daniel G. Bradley, "Y-chromosome Variation and Irish Origins," *Nature* 404: 351 (2000); Nicholas Wade, "Researchers Trace Roots of Irish and Wind Up in Spain," *New York Times*, March 23, 2000, p. A13; Nicholas Wade, "Y Chromosomes Sketch New Outline of British History," *New York Times*, May 27, 2003, p. F2.

315. James F. Wilson et al., "Genetic Evidence for Different Male and Female Roles During Cultural Transitions in the British Isles," *Proceedings of the National Academy of Sciences* 98:5078–5083 (2001).

316. Norman Davies, *The Isles*, Oxford University Press, 1999, p. 3.

317. Benedikt Hallgrimsson et al., "Composition of the Founding Population of Iceland: Biological Distance and Morphological Variation in Early Historic Atlantic Europe," *American Journal of Physical Anthropology* 124:257–274 (2004).

318. Agnar Helgason et al., "Estimating Scandinavian and Gaelic Ancestry in the Male Settlers of Iceland," *American Journal of Human Genetics* 67:697–717 (2000).

319. Agnar Helgason et al., "mtDNA and the Origin of the Icelanders: Deciphering Signals of Recent Population History," *American Journal of Human Genetics* 66:999–1016 (2000).

320. Benedikt Hallgrimsson et al., "Composition of the Founding Population of Iceland."

321. Nicholas Wade, "A Genomic Treasure Hunt May Be Striking Gold," *New York Times*, June 18, 2002, p. F1.

322. Agnar Helgason et al., "A Population-wide Coalescent Analysis of Icelandic Matrilineal and Patrilineal Genealogies: Evidence for a Faster Evolutionary Rate of mtDNA Lineages than Y Chromosomes," *American Journal of Human Genetics* 72:1370–1388 (2003).

323. M. F. Hammer et al., "Jewish and Middle Eastern Non-Jewish Populations Share a Common Pool of Y-chromosome Biallelic Haplotypes," *Proceedings of the National Academy of Sciences* 97:6769–6774 (2000).

324. Mark G. Thomas et al., "Founding Mothers of Jewish Communities: Geographically Separated Jewish Groups Were Independently Founded by Very Few Female Ancestors," *American Journal of Human Genetics* 70:1411–1420 (2002).

325. Harry Ostrer, "A Genetic Profile of Contemporary Jewish Populations," *Nature Review Genetics* 2:891–898 (2001).

326. Shaye J. D. Cohen, *The Beginnings of Jewishness*, University of California Press, 1999, p. 303.

327. Denise Grady, "Father Doesn't Always Know Best," *New York Times*, July 7, 1988, Section 4, p. 4.

328. Yaakov Kleiman, "The DNA Chain of Tradition," www.cohen-levi.org

329. Karl Skorecki et al., "Y Chromosomes of Jewish Priests," *Nature* 385:32 (1997).

330. Israel Finkelstein and Neil Asher Silberman, *The Bible Unearthed*, Free Press, 2001, p. 98.

331. Mark G. Thomas et al., "Origins of Old Testament Priests," *Nature* 394:138–139 (1998).

332. James S. Boster, Richard R. Hudson, and Steven J. C. Gaulin, "High Paternity Certainties of Jewish Priests," *American Anthropologist* 111(4):967–971 (1999).

333. Doron M. Behar et al., "Multiple Origins of Ashkenazi Levites: Y Chromosome Evidence for Both Near Eastern and European Ancestries," *American Journal of Human Genetics* 73:768–779 (2003).

334. Nicholas Wade, "Geneticists Report Finding Central Asian Link to Levites," *New York Times*, September 27, 2003, p. A2.

335. Harry Ostrer, "A Genetic Profile of Contemporary Jewish Populations."

336. Jared M. Diamond, "Jewish Lysosomes," *Nature* 368:291–292 (1994).

337. Neil Risch et al., "Geographic Distribution of Disease Mutations in the Ashkenazi Jewish Population Supports Genetic Drift over Selection," *American Journal of Human Genetics* 72:812–822 (2003).

338. Montgomery Slatkin, "A Population-Genetic Test of Founder Effects and Implications for Ashkenazi Jewish Diseases," *American Journal of Human Genetics* 75:282–293 (2004).

339. Gregory Cochran, Jason Hardy, and Henry Harpending, "Natural History of Ashkenazi Intelligence," *Journal of Biosocial Science*, in press (2005).

340. Melvin Konner, *Unsettled: An Anthropology of the Jews*, Viking Compass, 2003, p. 198.

341. Merrill D. Peterson, *The Jefferson Image in the American Mind*, Oxford University Press, 1960, p. 187.

342. Joseph J. Ellis, *American Sphinx: The Character of Thomas Jefferson*, Knopf, 1997.

343. Annette Gordon-Reed, *Thomas Jefferson and Sally Hemings: An American Controversy*, University of Virginia Press, 1997, p. 224.

344. Eugene A. Foster et al., "Jefferson Fathered Slave's Last Child," *Nature* 396:27–28 (1998).

345. Nicholas Wade, "Defenders of Jefferson Renew Attack on DNA Data Linking Him to Slave Child," *New York Times*, January 7, 1999, p. A20.

346. Edward O. Wilson, *Consilience*, Alfred A. Knopf 1998, p. 286.

347. The Chimpanzee Sequencing and Analysis Consortium, "Initial Sequence of the Chimpanzee Genome and Comparison with the Human Genome," *Nature* 436:69–87 (2005). The 99 percent agreement rests on a comparison of the human and chimp DNA sequences that directly correspond with one another. The human-chimp genome similarity falls to 96 percent after taking into account the insertions and deletions, i.e., stretches of DNA in one genome that have no counterpart in the other.

348. Louise Barrett, Robin Dunbar, and John Lycett, *Human Evolutionary Psychology*, Princeton University Press, 2002, p. 12.

349. Mark Pagel, in *Encyclopedia of Evolution*, p. 330.

350. Sarah A. Tishkoff et al., "Haplotype Diversity and Linkage Disequilibrium at Human G6PD: Recent Origin of Alleles That Confer Malarial Resistance," *Science* 293: 455–462 (2001).

351. J. Claiborne Stephens et al., "Dating the Origin of the CCR5–Δ32 AIDS-Resistance Allele by the Coalescence of Haplotypes," *American Journal of Human Genetics* 62:1507–1515 (1998).

352. Alison P. Galvani and Montgomery Slatkin, "Evaluating Plague and Smallpox as Historical Selective Pressures for the CCR5–Δ32 HIV-Resistance Allele," *Proceedings of the National Academy of Sciences*, 100:15276–15279 (2003).

353. Hreinn Stefansson et al., "A Common Inversion under Selection in Europeans," *Nature Genetics* 37:129–137 (2005); Nicholas Wade, "Scientists Find DNA Region That Affects Europeans' Fertility," *New York Times*, January 17, 2005, p. A12.

354. Yoav Gilad et al., "Human Specific Loss of Olfactory Receptor Genes," *Proceedings of the National Academy of Sciences* 100:3324–3327 (2003).

355. Patrick D. Evans et al., "Microcephalin, a Gene Regulating Brain Size, Continues to Evolve Adaptively in Humans," *Science* 309:1717–1720 (2005).

356. Nitzan Mekel-Bobrov et al., "Ongoing Adaptive Evolution of ASPM, a Brain Size Determinant in *Homo sapiens*," *Science* 309:1720–1722 (2005).

357. Li Zhisui, *The Private Life of Chairman Mao*, Random House, 1994.

358. Quoted in Bobbi S. Low, *Why Sex Matters*, Princeton University Press, 2000, p. 57.

359. Elizabeth A. D. Hammock and Larry J. Young, "Microsatellite Instability Generates Diversity in Brain and Sociobehavioral Traits," *Science* 308:1630–1634 (2005).

360. Richard E. Nisbett, *The Geography of Thought*, Free Press (2003).

361. Victor Davis Hanson, *Carnage and Culture*, Doubleday, 2001, p. 54.

362. Nicholas Wade, "Can It Be? The End of Evolution?" *New York Times*, August 24, 2003, Section 4, p. 1.

363. Bobbi S. Low, *Sex, Wealth, and Fertility*, in *Adaptation and Human Behavior*, edited by Lee Cronk, Napoleon Chagnon and William Irons, Walter de Gruyter, 2000, p. 340.

364. The reason is that before the generation of eggs or sperm, the chromosome inherited from the mother must align itself with the counterpart chromosome inherited from the father. For the alignment to take place successfully, the DNA of the two chromosomes must match fairly well throughout their length. If the chromosomes are too diverse, with too many different DNA units, they will not pair up correctly; viable sperm or eggs will not be created and the individual will be infertile. M. A. Jobling et al., *Human Evolutionary Genetics*, Garland, 2004, p. 434.

ACKNOWLEDGMENTS

This book grew out of conversations with the many scientists whom I interviewed in the course of writing articles about the genetics of human origins for the *New York Times*. I am most grateful to them for sharing their knowledge and insights.

I thank Peter Matson, of Sterling Lord Literistic, for shaping the idea of the book, and Emily Loose, its editor at Penguin Press, for shaping its structure. I am also grateful to the *New York Times* for giving me leave to write it.

Henry Harpending, an anthropologist at the University of Utah, Richard Klein, a paleoanthropologist at Stanford University, and Kari Stefansson of DeCode Genetics in Iceland, were kind enough to read parts of the manuscript. I thank them for their helpful emendations; they bear no responsibility for errors that may remain. I am indebted to friends who read the book in draft and saved me from many lapses, including Nancy Sterngold of Los Angeles, Jeremy J. Stone of Catalytic Diplomacy, Richard L. Tapper of the London School of Oriental and African Studies, and my wife, Mary V. Wade.

I thank the Velasco brothers of 5W Infographic for drawing many of the graphics and Steven E. Duenes of the *New York Times* for graphical advice.

CREDITS

FIGURE 2.1. Family tree of humans and other great apes. Adapted from M. A. Jobling et al., *Human Evolutionary Genetics*, Garland Publishing 2004, p. 222. Source: Pascal Gagneux et al., "Mitochondrial Sequences Show Diverse Evolutionary Histories of African Hominoids," *Proceedings of the National Academy of Sciences*, 96, 5077–5082, 1999, © 1999, National Academy of Sciences, U.S.A.

FIGURE 2.2. Evolution of human stone tool kits. From Richard Klein, *The Human Career*, University of Chicago Press, 2nd edition 1999, p. 576.

FIGURE 2.3. The three human species of 50,000 years ago. Illustration by 5W Infographic.

FIGURE 4.1. The universal human Y chromosome. Illustration by 5W Infographic.

FIGURE 4.2. The Y chromosome family tree and its geographical distribution. Illustration by 5W Infographic.

FIGURE 4.3. Mitochondrial DNA family tree and its geographical distribution. Illustration by 5W Infographic.

FIGURE 5.1. The route from Africa to the former continent of Sahul. Illustration by 5W Infographic.

FIGURE 5.2. The arrival of modern humans in Europe. From Paul Mellars, "The Impossible Coincidence. A Single-Species Model for the Origins of Modern Human Behavior in Europe," *Evolutionary Anthropology* 14,

12–27, 2005. © Paul Mellars 2005. Reprinted with permission of Wiley-Liss, a subsidiary of John Wiley & Sons, Inc.

FIGURE 6.1. The roller coaster of climate change in the Upper Paleolithic. Copyright Randall White, New York University.

FIGURE 6.2. The forced evacuation of Europe and Asia. Illustration by 5W Infographic.

FIGURE 7.1. The homeland of the Natufians, the first foragers to settle. From Ofer Bar-Josef, "On the Nature of Transitions: The Middle to Upper Palaeolithic and the Neolithic Revolution," *Cambridge Archaeological Journal*, 8:2, 1998, Cambridge University Press.

FIGURE 9.1 (in note 256) Genes that have undergone recent evolutionary change in the genomes of East Asians (ASN), Europeans (CEU), and Africans (YRI). From Benjamin Voight, Sridhar Kudaravalli, Xiaoquan Wen and Jonathan K. Pritchard, "A Map of Recent Positive Selection in the Human Genome," *PloS Biology* 4:446–458 (2006).

FIGURE 10.1. Large language families may have arisen through farming. Illustration by 5W Infographic, adapted with permission from Jared Diamond and Peter Bellwood, "Farmers and Their Languages: The First Expansions," *Science* 300, 597–603, 2003, © 2003 AAAS.

FIGURE 10.2. A geneticist's tree of the Indo-European language family. Illustration by Quentin Atkinson, University of Auckland, New Zealand.

FIGURE 10.3. The world's language superfamilies. From Merritt Ruhlen, *A Guide to the World's Languages*, p. 284–285. Stanford University Press 1991, © 1987 by the Board and Trustees of the Leland Stanford Jr. University.

FIGURE 10.4. The Afroasiatic language family. From Merritt Ruhlen, *A Guide to the World's Languages*, p. 284–285. Stanford University Press 1991, © 1987 by the Board and Trustees of the Leland Stanford Jr. University.

FIGURE 10.5. The distribution of Eurasiatic. From Joseph H. Greenberg, *Indo-European and Its Closest Relatives*, Stanford University Press, 2002, © 2000 by the Board and Trustees of the Leland Stanford Jr. University.

FIGURE 11.1. Thomas Jefferson's family with Sally Hemings. From M. A. Jobling et al., *Human Evolutionary Genetics*, Garland Publishing 2004, p. 492.

Index

Page numbers in *italics* refer to figure captions.